D0904887

SURVIVAL

STANLEY JOHNSON
& ROBERT VAGG

SURVIVAL

SAVING ENDANGERED MIGRATORY SPECIES

STACEY
INTERNATIONAL

Editor
Christopher Ind

Design
Graham Edwards

Stacey International
128 Kensington Church Street
London W8 4BH
Telephone: +44 (0)20 7221 7166
Fax: +44 (0)20 7792 9288
Email: info@stacey-international.co.uk
www.stacey-international.co.uk

ISBN: 978-1-906768-11-9

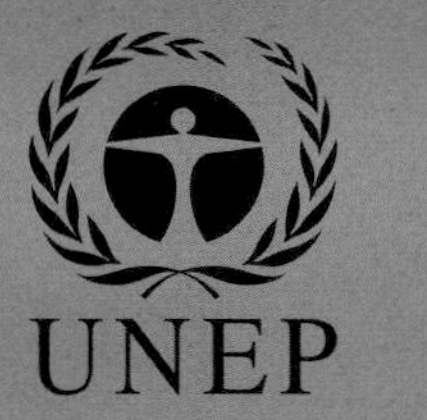

The authors are donating all royalties to the United Nations Environment Programme's Convention on Migratory Species (CMS). They gratefully acknowledge the support of both CMS and UNEP in making this project possible.

CIP Data: A catalogue record for this book is available from the British Library

Printed in China by 1010 Printing

Half title page:
The dugong (*Dugong dugon*), inhabitant of the deep in the Arabian Gulf, the Indian Ocean and the waters around northern Australia, is an ancient species feeding mainly on seagrasses.

Title page:
Probably the species that best symbolises the plight of wildlife in a warming world, the polar bear (*Ursus maritimus*) remains under threat from the breaking up of the ice sheets, though its numbers are thought to be stable.

Previous pages:
The previous pages carry a selection of images of endangered species on migration: African elephants (*Loxodonta africana*) arriving at new pastures; a pod of long-finned whales (*Balaenoptera physalus*) surfacing off the Norwegian coast; Blue wildebeest (*Connochaetes taurinus*) on the move through acacia woodland at sunset.

This page:
A lone cheetah (*Acinonyx jubatus*) surveys her prey as twilight falls.

Page 12:
Blue wildebeest (*Connochaetes taurinus*) and Zebra (*Equus quagga*) in Kenya's Masai Mara.

Page 24:
Snow geese (*Anser caerulescens*) flocking over New Mexico, USA.

Page 28:
Pink salmon (*Oncorhynchus gorbuscha*) pass Canada's Glen Cove en route for spawning.

Contents

Foreword

Migratory animals constitute those 8 to 10 thousand of the world's 1.8 million known species that travel at regular intervals, mostly between feeding and breeding grounds. Sometimes thousands of kilometres separate these areas but in other cases the distances are relatively short, straddling national boundaries.

Migratory species range in form from gorillas, leopards and antelopes to fish, turtles, bats and birds. They vary in size from whales and elephants to apparently frail and featherweight insects like the monarch butterfly. The factor uniting them all is that no one country can secure their survival on its own. International cooperation is an absolute prerequisite for their conservation and this adds further complexity and vulnerability.

This book, with some 200 photographs, seeks to explain some of the mysteries of migration and describe how we are addressing the ever-growing threats facing the amazing creatures which undertake these daunting journeys just to survive. Unfortunately the many naturally occurring hazards are being compounded by human-induced dangers – from direct hunting to bycatch, from barriers, such as dams, turbines and high voltage cables to ship strikes and from invasive species, inadvertently or deliberately introduced by people into new environments disturbing natural equilibrium, to the rapid changes to global climate.

As long ago as 1972, the nations of the world recognised the need for collaborative action at the Stockholm Conference on "the Human Environment". This gathering not only established the United Nations Environment Programme (UNEP) but also acknowledged the particular needs of migratory species. The Convention on Conservation of Migratory Species of Wild Animals (CMS) was the result: a vehicle through which the world's countries could achieve the necessary co-operation to conserve migratory species. The Convention's work was and remains a substantial challenge and, perhaps as a result, the first decade of the Convention in the 1980s was tough. It was human perseverance that built CMS and allowed it first to survive and then to flourish. Over the last 5 years, the Convention has increased the rate of progress very tangibly. The number of Parties has grown by almost 30% to 113, with 30 further countries as signatories to CMS Agreements. More importantly, the Convention has grown from a principally European body into a truly global one, with representation in Latin America, Africa, Asia and Oceania.

There are sections throughout the book explaining the important role played by Agreements and Memoranda of Understanding (MOUs) in delivering focused conservation effort. In recent years, the number of species Agreements and MoU has doubled from 12 to 24. These have come thick and fast in the last 3 years: Elephants, African and Pacific Cetaceans, Saiga Antelopes, South American Grassland birds and flamingos, Atlantic monk seals, Africa-Eurasian birds of prey, Indian Ocean and Pacific dugongs and, hopefully by the end of 2010, sharks and houbara bustards. A great achievement in 2007 was the negotiation of the prized Agreement for gorillas – not a soft, non-binding agreement, but an international treaty.

The steady progress we are making in recruiting smaller developing countries to the Convention is a sign that we are regarded as relevant to new objectives. Mozambique became the 112th party to the Convention in August 2009 and Ethiopia the 113th in January 2010. Similarly, we are making determined efforts to recruit Russia, China, Brazil (all three of which already participate in CMS Agreements) and a number of other key countries for animal migration.

In June 2009 CMS opened an office in Abu Dhabi – its first major centre outside Europe. This will oversee the new birds of prey and dugong agreements. It presents the Convention with a huge opportunity to engage more fully with the countries that surround the Indian Ocean. Through the generosity of the US Fish and Wildlife Service, the Convention has a presence in Washington, from where we can work day by day to create new partnerships.

We should not forget the ideals that inspired the founding fathers of CMS. It was to be a Convention for conservation – the only UN body specifically devoted to the conservation of species and of their ecosystems. It is fashionable today to assert that our environment must be protected and valued only for its economic potential. It is certainly an interesting approach, and there are more and more organisations addressing these aspects of the environment. But that is not what CMS is about – it is about the wonderful spectacle of animals and Nature and the compelling need to conserve them for the good of mankind. It is about the obligation we all have to ensure these treasures are protected for future generations. This is why CMS is a jewel among treaties.

I have had the privilege of working with both Stanley Johnson and Robert Vagg on wildlife work for a number of years. Johnson, himself a pioneer of legislation to protect European wildlife, is an active Ambassador for the Convention on Migratory Species, carrying our message far and wide, and giving us his vast experience of environmental policymaking and negotiation. Robert Vagg has single-handedly amassed a wealth of material, which he originally wrote for a CMS reference work. His commitment to CMS and migratory species over many years is also inspirational. I thank them both, and also UNEP for their financial grant, for without these elements this valuable new book could not have come into being.

Robert Hepworth,
Former Executive Secretary of CMS
March 2010

Opposite:
Populations of African elephants (*Loxodonta africana*) are showing signs of recovery after years of poaching for their precious ivory tusks.

Introduction by Stanley Johnson

In September 2009, as this book was being prepared for publication, I was able to spend a few days in the Masai Mara in Kenya. The Mara extends over 590 square miles, with an inner core of 250 square miles which is today designated as a National Reserve. Though I had been "on safari" here on several occasions over the last thirty-odd years, this was the first time I had witnessed the annual migration at its peak. It is sometimes called "the greatest wildlife spectacle on earth".

Seeing the animals from the basket of a hot-air balloon is an unbeatable experience. As you rise into the air, you gaze down at the vast expanse of plain below.

Below:
A hot air balloon provides the perfect vantage point for the author to view wild elephants in the Masai Mara.

As far as you can see, indeed right up to the Serengeti itself, the other side of the Tanzanian border, the grassy plains are literally black with animals. The sheer numbers involved are mind-boggling: over a million and a half wildebeest or gnus, half a million zebra, another half-million topis, elands and Thompson's gazelle.

With the sun behind us, the balloon cast a great shadow on the plains as we passed fifty or a hundred feet overhead. When the pilot fired the burner, the whoosh of igniting flame often caused a mini-stampede. Standing in the balloon's basket, we could hear the thunder of hooves and the squeals and rumbles of the herd. As we floated downwind, we seemed to open up a path in the sea of animals below, like Moses parting the waters of the Red Sea.

Later that day, back on terra firma, we watched in awe as hundreds of wildebeest and zebra gathered up the courage to cross the Mara River in the teeth of a small army of waiting crocodiles. We could see the animals, led – it seemed – by the zebras, coming down to the water, even taking a step or two across the rocks, then catching sight of the waiting crocodiles and quickly withdrawing to the safety of the bank, only to be jostled and harried by other animals hoping to cross.

Oddly enough, it was one lone zebra that broke the deadlock. By now, half a dozen crocodiles had nudged their way far upstream so that they were almost directly in the path of the migrating animals. With water levels so low, we could see virtually the whole length, breadth and height of the gigantic reptiles and if we could see them from where we were, the migrants certainly could.

But the little lone zebra seemed to have thought it out. He didn't try to dash past or even (heroically, on quick and dancing feet) over the crocodiles. Instead, he went downstream, round the back of them. An end-run, if ever there was one. Out of danger, he scampered up the bank.

That splendid solo effort was the signal for a sudden rush of animals. They came thick and fast, so thick and so fast, that it seemed that even the huge snapping jaws of the crocodiles were going to miss their mark.

The death we witnessed that morning by the Mara River had almost a balletic quality to it. This might be nature red in tooth and claw but still there was a terrible beauty about the way one crocodile managed to seize a young zebra, catching it by its throat and forcing its head under water, while three or four other crocodiles – hungry giants, all of them – swivelled into action in a stunning display of team-work.

If the Serengeti/Mara migration is indeed one of the world's most extraordinary wildlife spectacles, there are others which run it close. When I was a member of the European Parliament in the 1980s, I had the opportunity to visit the ice-floes in the Gulf of St Lawrence when the Harp seals were whelping. I flew over the frozen wastes in a helicopter and was stunned by the sheer mass of seals on the ice below. Some had already whelped and had their pups beside them; others had yet to whelp. How many were there? Opinions varied but it seemed likely that the number of Harp seals migrating at that time from the Arctic to Canadian waters numbered several million.

Later, we landed on the ice, leaving the engine running, and walked among the animals, awe-struck by the sheer scale and volume of this great biological event.

Another 'great migration' which I will never forget is that of the Grey whales off the coast of Baja California. Hunted virtually to extinction in the 19th and 20th century, the Grey whale has made an extraordinary recovery, and the population is now around 18,000.

Around 10am one morning, after waiting for the tide to rise, the small vessel which I had boarded in San Diego crossed the sandbar which separates San Ignacio lagoon from the open sea. Here each year, the Grey whales come to calve, the warm waters of the lagoon providing an ideal nursery for their young who, as it were, find their feet here before accompanying their mothers on the 6,000-mile return journey north to their feeding grounds in the Arctic.

Almost as soon as we had entered the lagoon, we could see whales spouting around us.

For a species that has absolutely no reason not to fear and loathe the human race, the Grey whale seems remarkably forgiving. Indeed, one of the remarkable features of whale watching in San Ignacio lagoon is that quite often this seems to be a two-way process. You can be out on the lagoon with a local boatman in one of the licensed pangas when a Grey whale, often with her calf, will push alongside the boat. They will raise their huge heads right over the side of the panga and you can find yourself, literally, eyeballing a 50-tonne monster, which could, if it so decided, send your frail craft to the bottom of the sea with one flick of its enormous tail.

I held out my hand to one animal as it approached us and felt the strange rubbery texture of the hide.

Our Mexican boatman that morning told us how a few years earlier, Mexico's then President Ernesto Zedillo came to the lagoon with his wife and family. This was a crucial moment. The Japanese giant Mitsubishi was pressing very hard for permission to open a huge salt factory on the lagoon that could have threatened the very survival of the Grey whale.

"The President and his wife and kids, they come out in my boat," the boatman told us. "The President's wife, she kissed the whale right on its head that day. I saw it. I was there. So the president, when he saw his wife kissing the whale, he said 'Right. No more salt factory. We keep the lagoon just for the whales.' And he announced the end of the salt project that very day!"

Above:
Herds of gazelle graze the grasslands of the Serengeti, where many of Africa's endangered migratory species are to be found – albeit in markedly fewer numbers than just 10 years ago.

A few years ago I spent a week on Playa Grande with the team from Earthwatch, the environmental charity, in Costa Rica's remote Guanacaste Province, trying to save the endangered Leatherback turtle. We must have walked 10 or 12 miles a night. We would start our patrol three hours before high tide and continue until three hours after, returning at daybreak – foot-sore and soaked to the skin – in time to catch a few hours' sleep.

For the past 25 years, men and women from all over the world have been volunteering to take part in this amazing endeavour. They have been out there, year in, year out, pacing the sands throughout the entire six-month nesting season of the Leatherback turtle, and – believe me – if they weren't there, the plight of this mighty animal would be even more precarious than it is.

The harsh truth is that, in the Pacific Ocean at least, the Leatherback turtle is critically endangered. Its nesting beaches, all around the Pacific Rim, have been turned into seaside resorts. If a female Leatherback does manage to reach the shore to lay her eggs, she may be hacked to pieces by waiting gangs, or her eggs, once laid, may be ruthlessly plundered. Industrial fishing, particularly long-lining, has further contributed to the tragic decline in leatherback numbers.

Playa Grande offers the last best hope of saving the species from extinction in the Pacific. Miraculously, the big-time developers have not yet got their claws into this part of the Costa Rican coastline. The bright lights from hotels and housing developments, so offputting to freshly hatched turtles, do not shine here. But, as I write these words, I learn that the area's protected status is under new and serious threat.

A quarter of a century ago, there were around 90,000 mating female Leatherbacks to be found in the Pacific. Today, there are fewer than 5,000. Witnessing this giant creature – which weighs almost a tonne – emerging at dead of night from the rolling Pacific sea-surge, hauling itself up onto the beach to lay its eggs, before heading back to the ocean, is one of the most stirring spectacles you are ever likely to experience. There is something primeval, elemental, about it.

Left:
A family of cheetahs (*Acinonyx jubatus*) remains ever on the alert. Cheetahs need a vast territory, and are especially vulnerable to man-made interference.

Since January 2007, I have been an honorary Ambassador for the United Nations Environment Programme's Convention on Migratory Species (CMS). Because the CMS deals with threatened migratory species, rather than the relatively abundant ones, my attention – and the focus of this book – has been on those 'threatened' species which are listed in the CMS appendices.

Take gorillas, for example. The CMS has recently sponsored a ten-nation treaty designed to protect the gorilla. Gorillas do not, of course, travel thousands of miles on an annual migration. But they can and do move from country to country, for example in the Virunga region where Uganda, Rwanda and the Democratic Republic of the Congo share a border. So there is a strong case for trying to develop, as the CMS has, the tools for international protection.

I saw the need for this at first hand in the Kahuzi-Biega national park in the eastern provinces of the Democratic Republic of the Congo, as well as in the Congolese Virungas.

We had set off from the park headquarters at Tsivanga at about 10am and spent the next two hours following a wildly gushing watercourse upstream, climbing steeply all the time. The trackers, as usual, were somewhere up there ahead of us and messages were passed regularly on the radio.

After a particularly strenuous uphill stretch, when it seemed that we were dragging ourselves up a vertical slope clutching at roots and branches, we heard a sudden stentorian roar as a fully grown male gorilla burst out of the undergrowth.

I knew what I was meant to do. The chief guide at Tsivanga, Robert Mulimbi, had briefed us. "If a gorilla charges, stand still," he said. "Lower your head. Look submissive." He looked pointedly at me. "Better wear a hat. If they see your fair hair, they may think you're another silverback."

Yes, I knew what to do all right. But when Chimanuka sprang from the bush in all his glory, I didn't stand my ground and lower my head. I jumped behind our pygmy tracker and held my breath.

This was a huge and magnificent animal. I had never seen anything like it before. We share 96 per cent of our DNA with gorillas. Man and gorilla may descend from a common ancestor.

Shock and awe. That's what you feel when you first see a gorilla in the wild.

Paradoxically, even though there are still more Grauer's gorillas in the world (and all of them in the DRC) than Mountain gorillas, the threat to the Grauer's may be more acute.

Take the eastern, more mountainous part of the Kahuzi-Biega National Park, the part we were in that day. In 1996, there were 254 gorillas there. Four years later, the number had fallen to 130. Today, there are probably less than 100. The continued presence of armed rebels in the park has been a major factor in this.

As far as the much larger western part of Kahuzi-Biega is concerned, the situation is even more dire. There are certainly substantial contingents of armed rebels inside the park. Another factor is the presence of as many as 8,000 "artisanal" coltan miners, mainly poor people who have made their way into the park to work the alluvial deposits of coltan or to quarry the minerals from the rocks.

As far as the gorillas are concerned, the combination of the two has been lethal. Nobody knows for sure how many Grauer's gorillas are left there. At one time, there were more than 10,000 in the lowland part of Kahuzi-Biega. Now the figure may be less than 1,000.

Last year I visited Mbeli Bai, a 37-acre (15-hectare) clearing in the heart of the Republic of Congo's equatorial rainforest, where the New York-based Wildlife Conservation Society (WCS) (which runs the Bronx and Central Park zoos) has been studying the life and habits of some 135 western lowland gorillas who visit the *bai*, or marshy clearing.

Mbeli Bai is remote. From Brazzaville, the capital, there is an hour's flight to Ouesso, followed by a five-hour ride in a motorized pirogue (dug-out canoe) to WCS headquarters at Bomassa. We were still on the river, with an hour of daylight left, when it began to rain. It rained harder and longer than I would have believed possible. When we finally disembarked, stiff and tired and soaked to the skin, we found that all the spare clothes we had were wringing wet – and didn't dry out for days.

But these are just the hazards of travelling in the Congo Basin. The great rainforests of central and western Africa contain one of the richest stores of biodiversity on earth, not least in the form of the western lowland gorilla which, in spite of the ravages of the Ebola virus, poachers and loggers, still exists in substantial numbers (possibly in the tens of thousands) in these regions. If ever there was a journey where the rewards far more than outweigh the hardships, the trip to Mbeli Bai is it.

As Thomas Breuer, the German scientist leading the WCS team at Mbeli, told me over a beer in camp one evening: "The bai, as you will have seen, has large deep pools of water among the swampy ground. One day I saw a female gorilla leave her baby behind at the edge of the pool while she went off to look for a stick to test the depth of water." Breuer's was the first-ever report of tool use by wild gorillas.

What's truly special here is that the scientists open up their mirador – the observation deck at the edge of the bai – to visitors, and share their knowledge of the gorillas with those who come here.

I do not wish to diminish the thrill of trekking gorillas in Rwanda or Uganda or even in the eastern Congo. But the special excitement of watching gorillas in Mbeli Bai in Congo Brazzaville is that these animals are unaware of observers. They are not "habituated", used to the close presence of tourists or researchers. At Mbeli Bai you are looking at one of the world's most extraordinary animals without any distorting filter.

The other reason I found gorilla-watching at Mbeli Bai – in the Nouabalé-Ndoki National Park – so satisfying was that there was no sense of the pressure of time. Most organised gorilla-trekking gives you an hour with the gorillas (if you are lucky enough to find them) and when that hour is up, you head back to camp. At Mbeli Bai, you have all day. At 8am on our first morning, we climbed up the steps of the mirador to find a group of gorillas already there. "That's Khan's group," Breuer told us before disappearing to his look-out station on the roof.

I looked Khan up in the handy catalogue. "Khan is an extremely large-bodied silverback with an enormous crest and monstrous neck muscle. Khan has tiny eyes, pointed ears and a large scar on the left part of his upper lip. His nostrils are large and pointed. Date of birth unknown."

As I watched the mighty animal, seated on his haunches less than 200 yards away, I reflected that the WCS catalogue entry was apt. If ever there was an example of raw, concentrated power, this was it.

There were 10 gorillas in the bai that first morning. Besides the great silverback, there were three adult females, two with infants, and a smattering of immatures and juveniles. Khan's group did not return during the rest of our time at Mbeli Bai. But on each of the three subsequent days, we saw a different group of gorillas, each with its own silverback leader. We saw almost 50 gorillas over four days.

This book deals with many of the migrating animals which are of concern to the CMS. I have had personal experience of only a fraction of that number. But many of the encounters I have had over the years, in addition to those mentioned above, have been truly memorable.

When, for example, I first met Roseline Beudels, one of CMS's Scientific Advisers, in Paris in September 2006, she told me why the CMS had decided to make Niger one of its priority targets.

Environmentalists over the past decade or so have, she explained, tended to concentrate on what they term "biodiversity hotspots", such as tropical rainforests with their extraordinary concentrations of fauna and flora. But the mandate of the CMS was to look after endangered migratory species wherever they were to be found, not just in the biological hotspots. And desert biodiversity, although less abundant in terms of number of species, is unique and most remarkable in terms of adaptation to extreme conditions.

Beudels told me about the CMS's project to prevent the Sahelo-Saharan antelopes from sliding into extinction. Six species altogether were covered by the CMS strategy: the Scimitar-horned oryx, the Addax, the Slender-horned gazelle, Cuvier's gazelle, the Dama gazelle and the Dorcas gazelle. The status of all these species, which had once been widespread throughout Saharan Africa, was now threatened or vulnerable. The scimitar-horned oryx had disappeared from the wild. The CMS was closely involved in a project to reintroduce the Addax in the wild in Tunisia, building on a captive herd which already existed in that country. As far as protecting the Addax in situ was concerned, Niger was a key country since it was thought to contain the last viable population of wild Addaxes. Between 100 and

Below:
The Mountain gorilla (*Gorilla beringei beringei*) migrates from higher-forested slopes in central Africa to bamboo forests for the few months that fresh shoots are available.

200 animals had in recent years been observed in the area around the Termit Massif and in the contiguous great desert erg known as Tin Toumma.

"The CMS," Beudels told me, "is determined to try to help Niger save the last wild Addaxes. We want to set up a protected area in around the Termit Massif and in Tin Toumma."

Six weeks later, I joined the CMS team in Niamey, Niger's dusty capital. Niger is one of the world's poorest countries. Each year the United Nations publishes a table called the Human Development Index. This is a comparative measure of life expectancy, literacy, education and standards of living for countries worldwide. Norway is top of the list. Niger – in 177th place – is at the very bottom, the lowest of the low.

It may at first sight seem perverse, in a country where human beings confront starvation on a daily basis, to talk about Niger's wildlife, but in reality protecting Niger's unique biological heritage is probably just as important in terms of basic socio-economic development as many of the other projects currently being undertaken.

The evening of my arrival in Niamey, Beudels, Greth and I met Ali Harouna, the Director of Niger's Department for the Protection of Wildlife. Harouna kindly drove out to see us at our hotel, outside the city centre.

While I fended off the mosquitoes, Harouna spoke of the need to involve the local people in the proposals to make Termit-Tin Toumma a Protected Area, were the project to stand any chance of success.

Above:
Alert wildebeest herd due to the presence of lions in the Masai Mara National Reserve, Kenya.

"The process of consultation may take a long time. In the end we will need a Presidential decree." He pointed out that, if you added a Termit-Tin Toumma Protected area to the existing protected areas in Niger, then almost 10 per cent of the country would be covered.

The key thing, of course, was not just to create another "paper park", but to have a system of protection that really worked on the ground.

The CMS team was able to confirm that the EU was likely to donate a substantial grant – more than €1.5m – to the Termit-Tin Toumma project, which was in addition to substantial funds already provided by the French Government's Global Environment Facility.

Harouna recognised that this was very good news. With the Scimitar-horned Oryx already extinct in the wild, saving the last viable population of Addaxes would be a tremendous coup for Niger.

The following Monday, in Zinder, a dusty town near the border with Chad which we reached after a 600-mile drive through the Sahel, the CMS team and the Niger Environment Ministry together inaugurated the Atelier de Lancement du Projet Antilopes Sahelo Sahariennes. Tribal chiefs and group leaders had already spent days travelling into town from the outlying areas. Now they had a chance to hear what the CMS proposed and to make their own comments.

For two days I sat at the back, looking over rows of turbanned heads, as one presentation followed another. The Tuaregs, the Toubou, the Hausa – all had their point of view and didn't hesitate to put it across. With prayer breaks as well as meal breaks to be taken, the whole event had a rather stately rhythm to it but, by the end, it looked as though the main objective had been secured.

Of course, the details still had to be sorted out: how big would the Protected Area be, how would a ban on hunting actually be enforced, how did you square Niger's evident determination to have a world-class Protected Area in Termit-Tin Toumma with the bizarre fact that some hunting concessions were still being granted, could there be teams of "eco-guards", what benefits would accrue to the local population? All these were important issues, but it seemed that at least the basic principles had been agreed.

I am sure that the fact that relatively large sums of money are going to be available to the project made a difference in the minds of the audience, but I believe there is more to it than that. I remember listening one morning to one of the tribal chiefs and being struck by the passion with which he spoke. He talked about how as a child he had grown up with wildlife. He had been to a nomad school and the gazelles would sometimes wander right up to the open-air class-room in the desert.

"La faune - c'est notre patrimoine!" he exclaimed. The applause from the other tribal leaders gathered there seemed both heartfelt and spontaneous.

After the workshop was over we left Zinder for Tesker. We then spent two days exploring the Termit Massif.

The Termit Massif is a most unusual geographical and biological feature. Extending almost 80 miles north to south and in parts more than eight miles wide, the rocky cliffs seem to rise hundreds of feet almost vertically from the desert floor. Here, if you are lucky, you will see Barbary sheep moving from crag to crag, desert tortoises, desert foxes and Dorcas gazelles. If you are very lucky, you might see a leopard or a Dama gazelle.

Of course, I was hoping desperately to see the rarest item of all, the Addax, even if that meant driving on east from Termit into the vast Tin Toumma desert erg. That Addaxes had been seen there in the past was not in doubt but the last sightings had been more than a year ago.

Next day our convoy moved on into the heat of the desert. We drove for several hours along a transect, our vehicles rising and falling with the sand-dunes. After 50km, we turned 90 degrees south for 10km, before returning on a track parallel to our original one.

I'd like to be able to record that at precisely that moment we had our first sighting of a herd of Addaxes, munching away on the unforgiving though still somehow nutritious desert grasses. But we had no such luck. The truth is that we were looking for a handful of animals in an area the size of Switzerland and it would have been almost a miracle if we had located them in such a short space of time.

The temperature in the desert dropped to 8°C that night and I was grateful for the shelter of my one-man tent. I lay with the flap open looking up at the stars.

Did it matter, I wondered, that we hadn't actually seen an Addax? Surely not. It was enough to know that somewhere in that vast desert, they are still there. And if the CMS project for a Termit-Tin Toumma Protected Area comes to fruition, as I have every reason to hope it will, there is a chance that the world's last remaining population of wild addax will not only survive but prosper well into the future.

This will be good for the Addax. And it will be a triumph for Niger as well.

I have mentioned some encounters with marine and terrestrial species. I have not done the count but I suspect that there are more migratory bird species than for either of the other two categories. In Bhutan last year, my wife and I had the chance to see the grey-necked crane on its winter feeding grounds. Though I am not a dedicated 'twitcher', I knew enough about the status of the highly-endangered species to be aware that we were privileged indeed.

I have had the good fortune to visit Antarctica twice so far: in 1984 and again in 2007. On the more recent trip I became aware that we seemed to be encountering fewer albatrosses than had previously been the case. I did not on this occasion visit, as I had in 1984, the great albatross breeding colonies in South Georgia, Bird Island or the Falklands/Islas Malvinas. But I did have a chance to talk to experts, both on board our vessel and subsequently in the UK. The consensus view seemed to be that, yes, there were far fewer albatrosses around.

The good news seemed to be that the level of international concern for the plight of the albatross was rising rapidly. Several of the people I spoke to suggested that if I really wanted to find out what was going on I should head to Hobart, capital of the Australian island of Tasmania.

"That's where this story is playing out," my old friend Rob Hepworth told me. Hepworth is the former Executive Secretary of the United Nations Convention on Migratory Species and he has written the Foreword to this book. The Convention encourages the setting up of special arrangements for the conservation of various threatened species. In that context, an international Agreement on the Conservation of Albatrosses and Petrels (ACAP) had recently been signed and ratified by more than a dozen states with interests in the Southern oceans. The headquarters of ACAP was in Tasmania, Australia.

So I headed for Hobart and, the day after my arrival, went to ACAP's office in Salamanca Square. Warren Papworth, a 50-year-old

Above:
The Addax (*Addax nasomaculatus*), a member of the gazelle family is one of Africa's most persecuted large mammals. A recent increase in loss to its habitat has depleted numbers yet further.

Australian who serves as ACAP's Executive Secretary, reviewed the global status of albatross populations for me.

"There are 19 albatross species currently listed by ACAP, 15 of which are classified as threatened with extinction," he told me. "Four species are currently recognized as 'critically endangered', five are 'endangered', six are 'vulnerable' and four are 'near-threatened'."

We were joined by Dr Rosemary Gales, a marine scientist working for the Tasmanian department of Primary Industries and Water, who serves as Convenor of Acap's Status and Trends Working Group. "Albatrosses are among the most threatened species of birds in the world," she told me, flashing up a series of slides to make the point. I learned that as long ago as 1993, Dr Gales had reviewed the global status of albatross populations and the factors affecting them. She had concluded that mortality associated with commercial fishing operations had become the most serious threat facing these birds.

The key issue then was "Albatross bycatch", in which the birds are caught by longlines targeting tuna, toothfish and other species. More recently, trawl fisheries have also been identified as a major threat, the birds being attracted by discarded fish and offal only to be caught on trawl cables and drowned.

If I learned that morning about the gravity of the situation in conservation terms, I also learned about the important steps being taken to address the problem. "Above all, we have to work with the fisheries management organisations," Papworth told me. So much depends on the willingness of the fisheries to cooperate. In the case of trawl fisheries, as for the longline fisheries, there are technical solutions, which can make a huge difference to the survival of albatrosses and other sea birds. If the fishing vessel, for example, can arrange its affairs so that it separates the discard phase from the catch phase, the number of birds clustering round the stern of the fishing vessel and becoming entangled in the warps can be dramatically reduced.

While I was in Tasmania, I also talked to Graham Robertson, who works for the Australian Government's Antarctic division and who for some years has been a leading figure in the effort to develop "mitigation" measures. As far as the longline fisheries are concerned, they revolve around the way in which the hooks are baited, the rate at which the line sinks, and the effectiveness of counter-measures, such as 'tori' streamers which are designed to deter birds.

Graham Robertson believed that as far as sea-bed (or "demersal") fisheries were concerned, important progress had been made, at least within the framework of the Convention for the Conservation of Antarctic Marine Living Resources (CCAMLR). This is a treaty with teeth. Mitigation measures can be mandated and enforced, although the problem of illegal, unreported and unregulated fishing remains. Another problem is that albatrosses range far and wide and certainly into latitudes where CCAMLR does not run. That is why the work of Acap and its ability, actual or potential, to influence fisheries-management organisations is so vital.

Apart from Acap, CCAMLR and the Australian Antarctic division, there is another major Hobart-based plank in the international effort to save the albatross: the Albatross Task Force (ATF), set up under Birdlife International's seabird programme. I had dinner one night with its co-ordinator, Ben Sullivan, who told me that one of the Task Force's main objectives is to increase the number of observers on fishing vessels. "The long-term solution," he said, "is not to discharge offal when you're trawling. We try to put our people on board the boats themselves. Actually, we don't call ourselves observers, we call ourselves instructors. We are two years into the programme now. We try to work with the fishing industry, but we need the support of government too."

Even if you do solve the problems associated with fisheries – and it's a big "if" – there are still the other threats to deal with. Gough Island, for example, a United Kingdom Overseas Territory in the South Atlantic, is home to the Tristan albatross. This species is now critically endangered, not by fishing, but by "supermice". These are ordinary house mice which arrived on the island on sailing ships decades ago, and have developed into monster rodents which burrow into the albatrosses' nests and eat the hapless chicks while they are still alive.

Since Britain's Foreign and Commonwealth Office is directly responsible for Gough Island, one can only hope someone puts a note on the Foreign Secretary's desk with an urgent sticker on it saying: 'Action this day!'

Finally, a word about sharks, highly migratory species, which have had a bad press ever since the film *Jaws* was released, if not before. The Great white shark which cleared the beaches of Amity Island, Massachusetts, one long hot summer thirty-eight years ago, has taken on an almost Freudian significance, like Moby Dick, the great white whale.

The sight of a fin in the water, any fin, has people stampeding for the safety of the shore. From a conservationist point of view, it's hard to underestimate the damage done by just one no-doubt well-intentioned film director (Steven Spielberg) based on the work of just one no-doubt well-intentioned novelist (Peter Benchley).

You can talk about the need to save pandas, polar bears, elephants, turtles, even great crested newts, without losing your audience. Try to tell people that the threat to the world's sharks is one of the most important wildlife issues confronting us today and the odds are they will, at the very least, look at you as though you need your head examined.

And yet, in reality, sharks are under attack as never before. They are being targeted by fisheries all over the world.

Their fins are cut off while they are still alive, then they are thrown back into the ocean to suffer a slow painful death by drowning. You don't have to travel to China to find shark-fin soup on the menu. Try Soho in London. Try Greenwich Village in New York. Try just about anywhere.

The statistics are horrifying. According to the United Nations Food and Agriculture Organisation (FAO), global shark catch reached 880,000 tonnes in 2003, an increase of 17 per cent on the level recorded just a decade earlier. And the odds are that these figures are substantial underestimates since they do not include discards.

Left:
Whale sharks (*Rhincodon typus*) are capable of diving to depths of 700m (see pages 76-77)

The reality is that world-wide fisheries directed deliberately at sharks, together with the death and destruction of sharks as a result of bycatch (you're aiming to catch tuna, but you catch a shark instead or as well), has resulted in a situation where, today, almost 50 per cent of the world's migratory shark species are classified as 'critically endangered', 'endangered' or 'vulnerable', with almost 30 per cent being classified as 'near threatened'.

The fishing fleets of around 20 nations account for 80 per cent of the world's shark catch. Taking the period 1990-2004, the top scorer was Indonesia with 12.3% of the world's catch, followed by India (9.1%), Pakistan (5.8%), and Spain (5.7%). The United States comes in seventh with 4.6%. and France 11th at 2.9%. The UK features in the list (at 13th), as does Portugal (16th).

Indeed, since fisheries is a competence of the European Union as such, rather than the individual EU nations, we have to recognize the harsh fact that the EU as a whole has been one of the principal agents of world-wide shark destruction.

Our fleets have been out there in international as well as European waters and sharks in the Atlantic, Pacific and Indian Oceans have paid the price.

Admittedly, there have over recent years been some cosmetic approaches to address the problem. The UN General Assembly has called for a shark finning ban and some fishing nations, and at least one of the regional fisheries management organisations (RFMOs), have responded. But you have to look at the small print. This is an Alice-in-Wonderland world. A shark finning ban is not the same as a ban on shark finning. Complex calculations relating to the weight of fins and landed carcass means that in practice fins can still be removed from living sharks and that finning bans in any case cannot be seen as an effective tool to control shark mortality.

As far as the latter point is concerned, the FAO adopted in 1999 an international plan of action on sharks, but that seems to have been honoured more in the breach than the observance.

For three species of shark – namely the great white (of *Jaws* fame), the basking shark and the whale shark – the situation is improving because they have been listed both by the Convention on International Trade in Endangered Species of Fauna and Flora (CITES) as well as by the Convention on Migratory Species (CMS). This is progress but it does not go nearly far enough – ironically, however spectacular these three so far listed species may be, they are not as seriously threatened as, for example, the Daggernose or Angel shark which are both listed as 'Critically Endangered'.

Conservationists have been arguing for years that what is needed now is real, as opposed to cosmetic, bans on finning, together with moves by the fishing nations and the RFMOs to ban completely the taking of the most seriously threatened shark species, together with the setting of strict catch limits for less-threatened but still vulnerable species.

At last, that call seems to have been heeded. In December 2007, the Government of the Seychelles hosted a meeting of more than 40 governments, together with representatives of fisheries bodies and non-governmental organisations. Those present agreed that the time had come, finally, to put in place an international mechanism to protect the world's sharks. Such a mechanism would specify not only the species of shark to be included but also the measures to be taken.

My own view is that CMS decisions on sharks should be binding in law and enforceable in practice. Of course, the detailed application of the agreement should be left to the competent authorities at national level, working as necessary through the regional fisheries organisations. But at least a clear framework for conservation and management would have been set.

Sharks have been around for 400 million years, longer than most other species, a fact which makes the prospect of their imminent extinction particularly poignant. Maybe, after the Seychelles meeting and the follow-up meeting which was held in Rome in December 2008, things are at last looking up for sharks. Watch this space!

Stanley Johnson, March 2010

Introduction by Robert Vagg

The origins of the Convention on the Conservation of Migratory Species of Wild Animals (CMS) can be traced back to the UN Conference on the Human Environment, which was held in Stockholm in June 1972. The main outcome of the Conference was the decision to establish the United Nations Environment Programme (UNEP), but delegates also recognised the unique needs of the world's endangered migratory species and agreed that an international instrument to protect them and mitigate the threats they faced was necessary.

The Government of the Federal Republic of Germany accepted the mandate to lead the negotiations to conclude an international treaty to conserve migratory animals. With the support of the IUCN's (International Union for Conseration of Nature) Environmental Law Centre, officials of what is now the Federal Ministry of the Environment, Nature Protection and Nuclear Safety first set about composing a draft text. After exhaustive consultations with other governments, an international conference was convened to negotiate the text, which was finally agreed and signed on 23 June 1979. It is because this final negotiation meeting was held at Bad Godesberg in Bonn that CMS is also known as the Bonn Convention.

Because their habitat requirements are greater, migratory animal species are generally more at risk of becoming endangered than sedentary ones. They need breeding grounds for reproduction and to raise their young; they need different sites to spend the winter; and they also need staging posts along their migration routes. All of these sites must provide food and shelter and if any link of this chain breaks, the chances of the animal's survival are greatly diminished. As well as facing natural barriers and hazards such as mountain ranges, deserts, oceans and predators, migratory species have to overcome further obstacles thrown in their way by mankind. Many human activities are detrimental to wildlife: our insatiable demand for natural resources has led to the loss of natural habitats, atmospheric pollution and unsustainable hunting and fishing practices. Bycatch – the incidental capture of non-target species in fishing nets – is a common threat facing marine turtles, sharks, dolphins and albatrosses.

The fact that migratory species cross international jurisdictional boundaries – artificial political constructs often without any ecological relevance – means that conservation measures can only be effective if undertaken multilaterally among all the Range States crossed by the animals during their migration. This collective obligation is recognised in the text of the Convention, which states: "Wild animals in their innumerable forms are an irreplaceable part of the Earth's natural system which must be conserved for the good of mankind".

The Convention has two Appendices. The first lists migratory species that are endangered and for these the Parties must conserve and restore habitats, mitigate obstacles preventing migration and take measures to reduce and eliminate the factors leading to the species being endangered. Appendix II lists migratory species which have an unfavourable conservation status and require international agreements for their conservation and management, as well as those which have a conservation status which would significantly benefit from the international cooperation that could be achieved through CMS. The Convention allows a degree of flexibility in that different populations of a species can be handled separately. Birds and mammals make up the majority of the species listed, but there are some reptiles (such as marine turtles), fish (such as sharks and sturgeons) and even one insect, the Monarch butterfly.

CMS is a framework or umbrella Convention. It is implemented primarily through species-specific, regional instruments. Some of these are Agreements, legally binding international treaties in their own right, with their own Secretariats and budgets, while others are less formal, non-legally binding Memoranda of Understanding (MoU). In CMS's thirty years, twenty-four such instruments have come into being – seven binding Agreements and seventeen MoU, with more in the pipeline. The species covered range from Wadden Sea seals to European bats, from albatrosses and petrels to the Bukhara deer and Saiga antelope. The MoU on the ruddy-headed goose has just two Parties – Argentina and Chile – while the Gorilla Agreement covers all ten Range States in Central Africa. Negotiations are under way for the first CMS instrument to address fish conservation, a global agreement concerning migratory sharks.

Since the conclusion of the negotiations in 1979 and its entry into force in 1983, the Convention has grown considerably. There were just fifteen Parties at the outset, with the 50th acceding in 1997 and the 100th in 2007. Mozambique brought the total to 112 in August 2009 and a further thirty-two non-Parties participate in the various Agreements and MoU. All regions of the world are now represented, with participation highest among the European and African countries, while recent growth has been strongest in Latin America. The Secretariat, which has been located in Bonn since it was established in 1984, is now housed in the UN Campus on the banks of the Rhine in the former Government Quarter.

CMS has a distinct and vital role to play in ensuring the survival of migratory species: a unique element of our shared natural heritage.

Robert Vagg, March 2010

Opposite:
The Bactrian deer (*Cervus elaphus bactrianus*), a sub-species of the European red deer, inhabits low-lying forests in Central Asia.

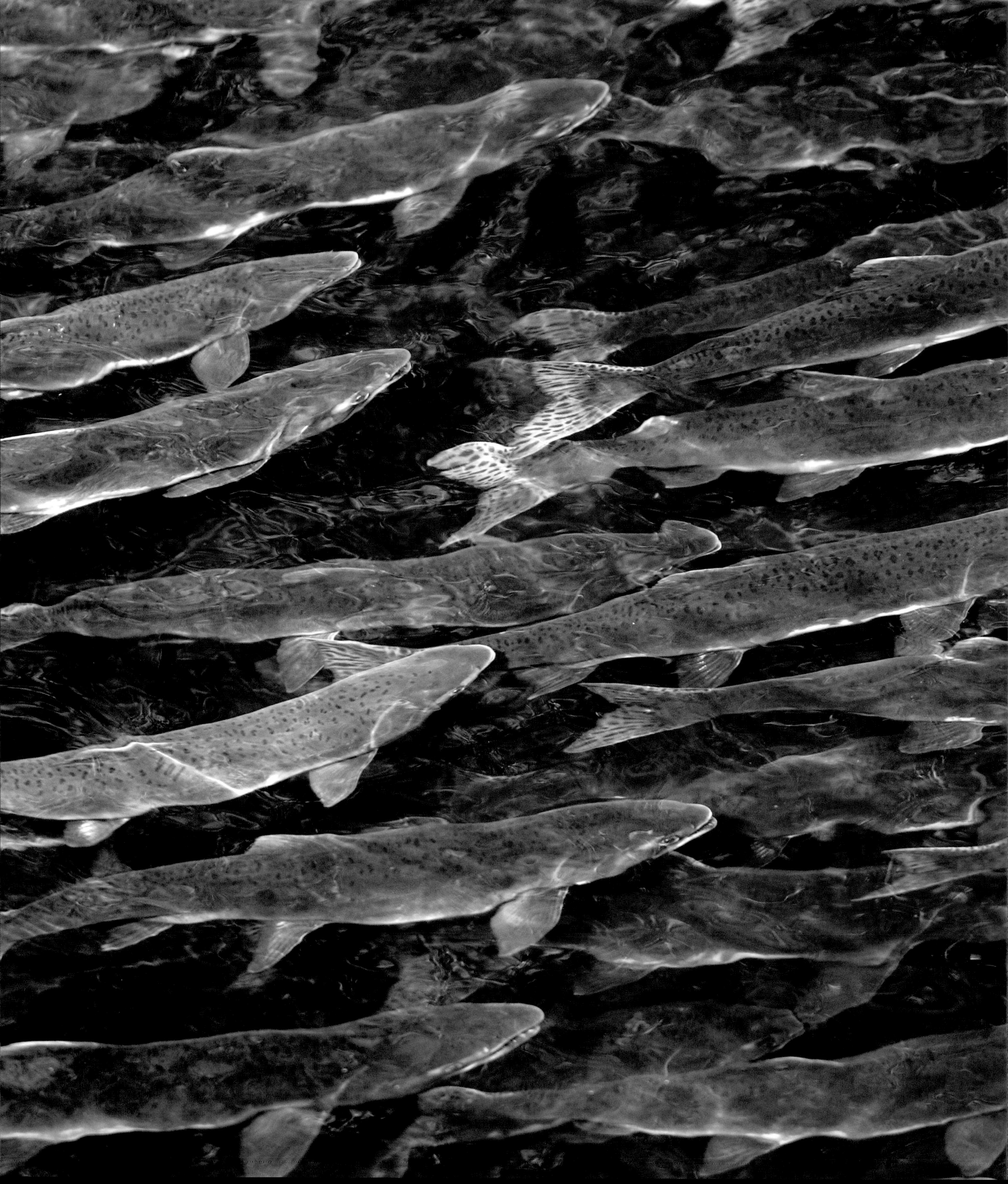

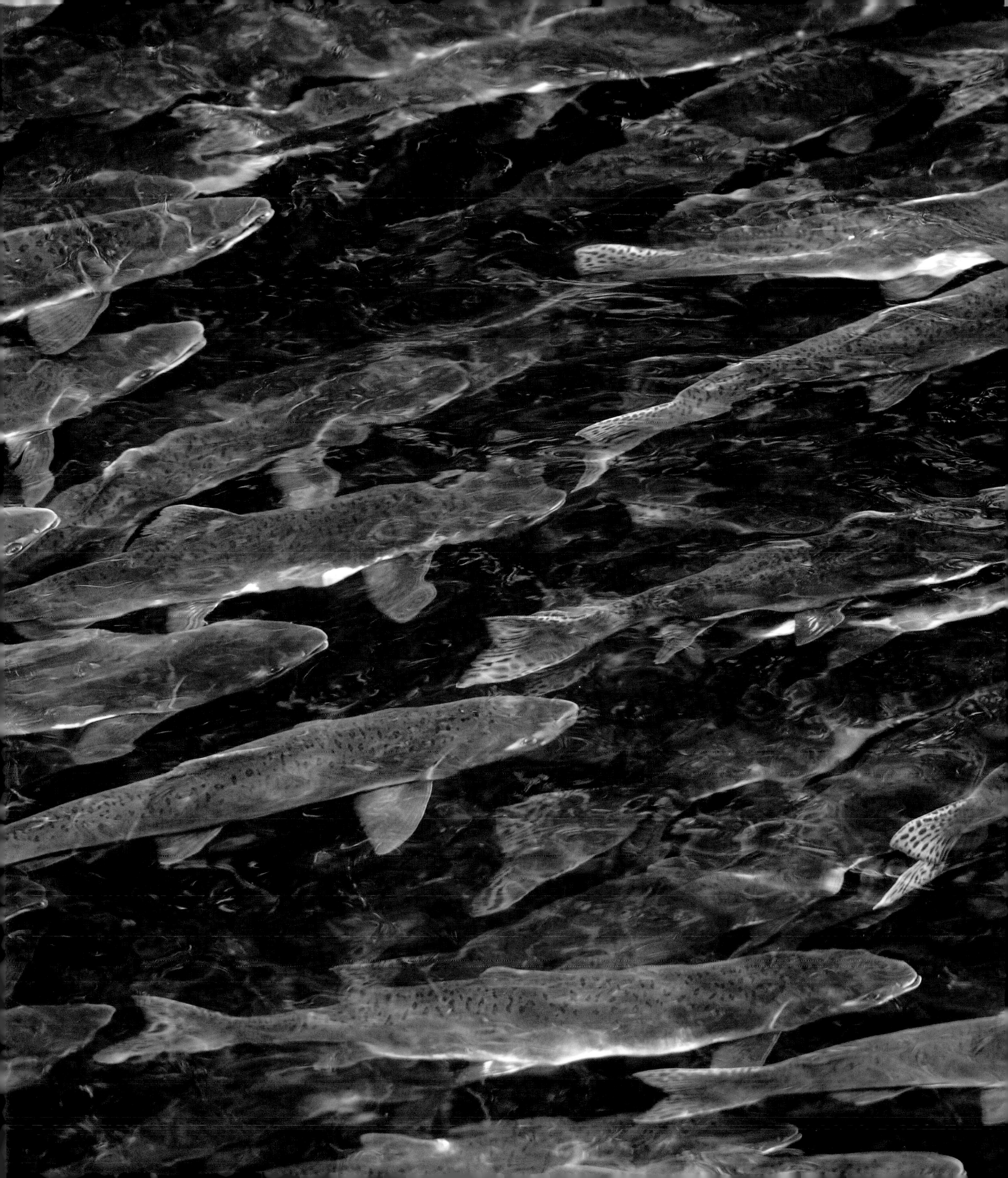

1

BIRDS

Surely no other group in the animal world better defines the free spirit of migration than that of the birds. Many species travel half the world, twice yearly, to breed where food is plentiful and overwinter in warmer climes. Some swallows famously migrate from the Arctic circle to the Antarctic, always returning to the same nesting site.

For any species, the route is fraught with danger. If they are not shot or eaten they risk exhaustion or being blown off course by freak weather.

In addition, global warming is dramatically altering the migration routes of many birds, further heightening the threat to their survival, as they are forced to undertake ever more arduous journeys.

MIGRATORY BIRDS

FROM THE ARCTIC TO THE ANTARCTIC

Due to their often spectacular and long-distance journeys, birds are perhaps the best-known group of migratory animals. Many species migrate from high latitudes to the tropics and beyond. One species, the Arctic Tern, an elegant white seabird, even breeds in the Arctic and migrates to the Antarctic!

To be able to complete their life cycles successfully, migratory birds not only need their natural breeding and wintering habitats to be preserved; they also require their traditional travel routes ("flyways") and stopover sites to be maintained. This makes it particularly challenging to conserve migratory birds.

In April 2009 CMS published *A Bird's Eye View on Flyways*, bringing together the key facts about migratory birds, their populations, the main flyways they use, the threats they face along the way, the benefits they bring to people and the environment and the actions we must take to help them survive, which include strengthening the international framework through CMS and its related agreements.

Bird migration spans vast areas comprising thousands of kilometres over land and sea and covering many countries. The birds play a crucial economic and ecological role affecting the millions of people who live along the migration routes. These journeys, perilous at the best of times, given that they involve crossing oceans, mountains and deserts, are made even more hazardous by human interventions.

Modern infrastructure is obstructing the flight pathways, and critical sites required for breeding, feeding, resting or moulting have shrunk in size and number to become islands in our human landscape. Indiscriminate hunting, which is particularly detrimental at key migration locations, and more recently climate change, have put populations of migratory birds under ever increasing pressure. Almost one fifth of the world's extant bird species are considered ecologically migratory. Of these more than a tenth are categorised as threatened or near-threatened by extinction according to the IUCN Red List. Migratory raptors in the Africa-Eurasia region are particularly threatened; just over half of the species listed have an unfavourable conservation status.

The value of migratory birds is often underestimated. This includes their role in providing ecological services as well as direct economic benefits. High income is generated from people who enjoy watching birds during their migrations. The annual income generated by just four examples of tourist birdwatching in North & Central America and South Africa is estimated at over US$13 billion, largely through sales of equipment. The monetary, environmental and spiritual wealth generated by migratory birds increases the incentive to conserve them. Virtually all countries of the world share a responsibility to halt the decline in numbers of affected avian species and the degradation of their habitats. Combating such threats can be best achieved by looking at bird migration in a broad context and by undertaking conservation work in a structured manner, along the entire extent of their flyways.

Conserving single bird species

In addition, some of the rarest birds in the world are covered by CMS regional Agreements in the form of Memoranda of Understanding. With the support of the Convention, the tiny populations of the Siberian crane are benefiting from captive breeding and the releases of young birds, which are taught their traditional migration routes by hang-glider pilots. The Siberian Crane Wetland Project, built on the MoU, forms the next step of the long-term programme to secure the species' survival. The Slender-Billed Curlew, one of the rarest of all migrants, is the subject of urgent efforts under CMS to discover its last winter refuges and where it breeds in the vastness of Eurasia. The main challenge of conserving the spectacular Great Bustard under a CMS Memorandum of Understanding is to manage modern agriculture throughout the bird's range in Central Europe.

Another species covered by a CMS MoU, the Aquatic Warbler, is a small songbird, which totally depends on a dwindling number of sites of a particular wetland type in Europe. Fortunately, the great majority of the key Range States has now signed up to save it, using the instruments of the Convention.

Conservation across continents

A major agreement for the conservation of waterbirds of the Central Asian flyway is envisaged under the framework provided by CMS. The Agreement on the Conservation of Albatrosses and Petrels (ACAP) tackles threats to these ocean wanderers, which range from drowning on long-line hooks set by commercial fishing vessels, to the taking of eggs and young birds by cats and rats. Birds of prey in the African-Eurasian region have a poor conservation status. They are subject to a variety of human induced threats, such as habitat loss or degradation, hunting, illegal shooting and poisoning. Collisions with aerial structures and electrocution by power lines also contribute to population declines. The recently-signed agreement for migratory birds of prey, increasing the profile of owls and raptors in the region, will promote more effective conservation through the international co-ordination of action tackling the threats to migratory birds of prey.

Birds in the western hemisphere

In South America, a Memorandum of Understanding for the two endemic species of Andean Flamingo, living in high-altitude lagoons subject to intense human pressure, has recently been negotiated. These flamingos migrate in the wetlands to forage and depend on the conservation of these habitats. Human activities such as agriculture, mining and unregulated tourism have been the main reason for the drop in population size. Southern South American Migratory Grassland Bird Species and their habitats are the focus of another Memorandum of

Understanding. Fragmentation of grassland and illegal capture and trade have been the main reasons for the populations' decline. The aim of the Action Plan is the protection of the habitats and the birds in Argentina, Brazil, Paraguay, Bolivia and Uruguay.

The Ruddy-Headed Goose has been persecuted especially in its wintering grounds in the South of the Argentinian province of Buenos Aires. This exclusively South American Memorandum of Understanding has been concluded between Chile and Argentina to save this population from the imminent danger of extinction.

The Agreement on the Conservation of African-Eurasian Migratory Waterbirds (AEWA)

The Agreement on the Conservation of African-Eurasian Migratory Waterbirds (AEWA) is the largest of its kind developed so far under CMS. It was concluded on 16th June 1995 in The Hague, the Netherlands, and entered into force on 1st November 1999 after the required number of at least fourteen Range States, comprising seven from Africa and seven from Eurasia had ratified. The Agreement is an independent international treaty.

The AEWA covers 255 species of birds ecologically dependent on wetlands for at least part of their annual cycle, including many species of divers, grebes, gannets, pelicans, cormorants, herons, egrets, bitterns, storks, crakes, flufftails, coots, moorhens, plovers, pranticoles, the black-winged stilt, the pied avocet, oystercatchers, rails, lapwings, snipes, godwits, curlews, redshanks, greenshanks, sandpipers, knots, ibises, spoonbills, flamingos, ducks, swans, geese, cranes, waders, gulls, terns and even the southern African penguin. Among the threats are habitat destruction and degradation, depletion of food supplies (through overfishing for example), unsustainable levels of poaching and climate change.

The agreement covers 118 countries and the European Community (EC) from Europe, parts of Asia and Canada, the Middle East and Africa. In fact, the geographical area covered by the AEWA stretches from the northern reaches of Canada and the Russian Federation to the southernmost tip of Africa. The Agreement provides for coordinated and concerted action to be taken by the Range States throughout the migration system of waterbirds to which it applies. Of the 118 Range States and the EC currently 62 countries (as of 1st November 2008) have become a Contracting Party to AEWA.

Parties to the Agreement are called upon to engage in a wide range of conservation actions which are described in a comprehensive Action Plan. This detailed plan addresses such key issues as: species and habitat conservation; management of human activities; research and monitoring; education and information; and implementation. An important element of the strategy is the development of international action plans.

Previous Pages:
The Arctic Tern (*Sterna paradisaea*), pictured here over Canada, has one of the longest migratory routes of all species.

Right:
Map to show the flyways of migratory bird species worldwide.

Overleaf:
A flock of Barnacle Geese (*Branta leucopsis*) migrates to its wintering grounds in western Scotland.

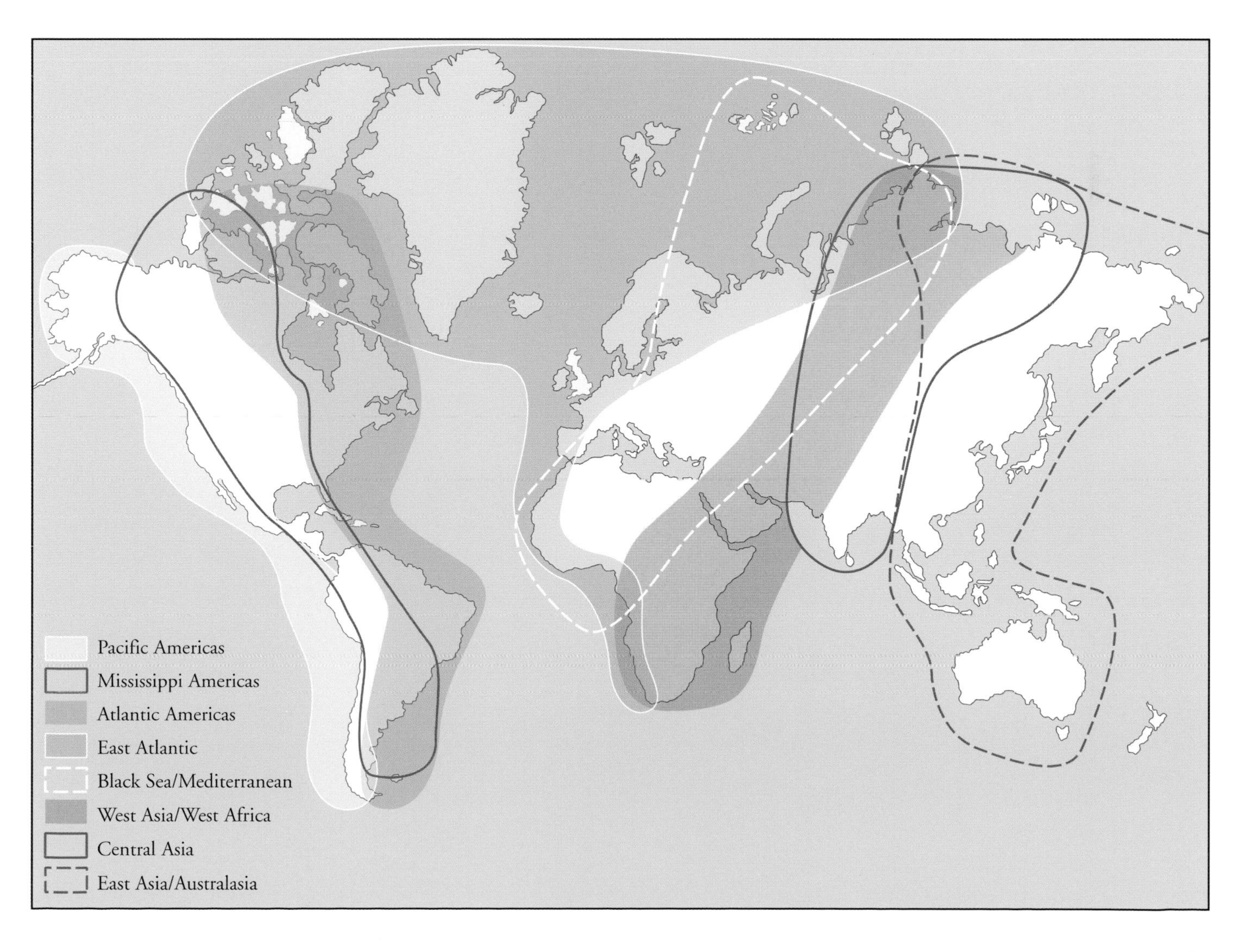

BRENT GOOSE

Branta bernicla

The Brent goose has a very large range in the northern hemisphere. It occurs in Belgium, Canada, China, Denmark, France, Germany, Iceland, Ireland, Japan, the Netherlands, Norway, Spain, Russia, the United Kingdom and the USA. Its preferred habitat is the seashore.

The global population of the Brent goose is estimated to be 570,000 individuals. Three subspecies are known: the Black, or Pacific Brent goose (*Branta bernicla nigricans*), the Dark-Bellied Brent goose (*Branta bernicla bernicla*) and the Light-Bellied Brent goose (*Branta bernicla hrota*). The Black Brent goose breeds in eastern Siberia, Alaska and northern Canada and migrates in the winter to the Asiatic and North American Pacific Coast. The Dark-Bellied Brent goose breeds in northern Siberia and winters in north-west Europe, and the Light-bellied Brent goose breeds in eastern Canada, Greenland and Svalbard and winters on the Atlantic coast of North America and in north-west Europe.

Hunting and climate change pose significant threats to the Brent goose. The Brent goose will face greater competition from species with a newly expanded or shifted distribution range if climate change takes hold. The effect of this additional threat has yet to be seen.

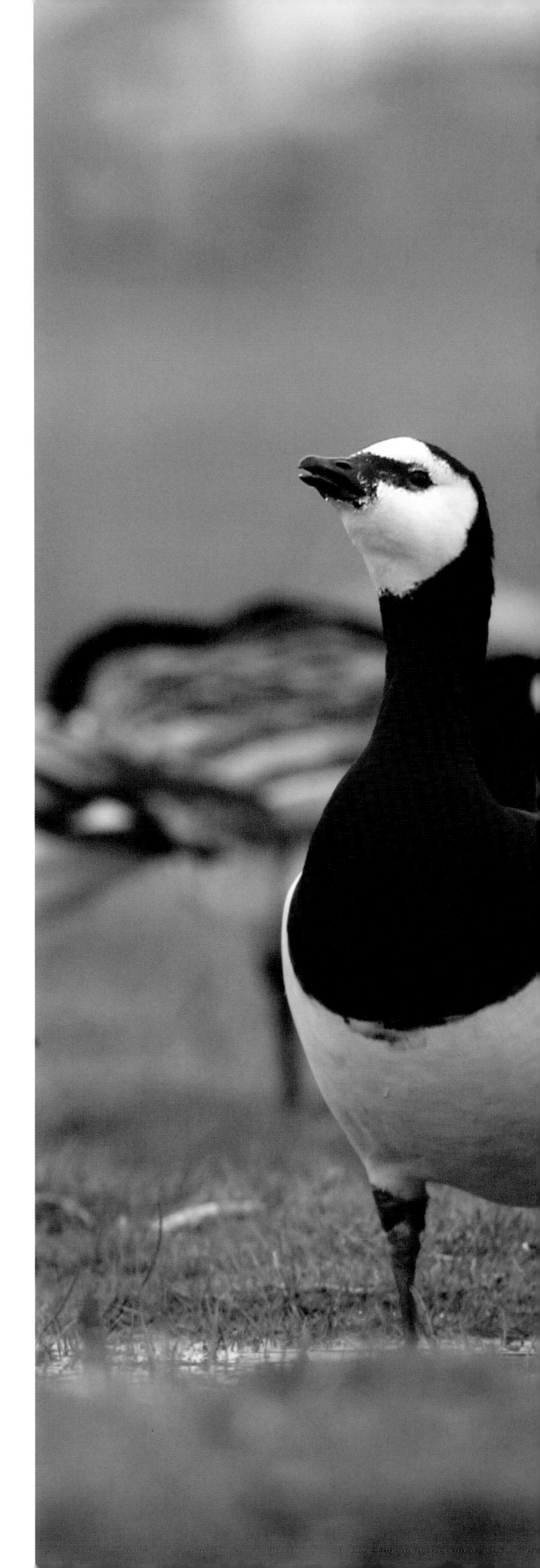

OPPOSITE:

SOCIABLE LAPWING

Vanellus gregarius

The Sociable lapwing or Sociable plover occurs across Central Asia and the Middle East, particularly on the Arabian Peninsula. It breeds in dry steppes, semi-deserts and grazed areas with low vegetation cover, and winters on dry plains, sandy wastes and short-grass areas adjacent to water.

The Sociable lapwing has been declining since the beginning of the twentieth century. The population is estimated at 600 to 1,800 birds. In February they start migrating from the wintering areas in Israel, Eritrea, Oman, Pakistan and India and arrive at their breeding grounds in south-central Russia and Kazakhstan in May and June. They leave there in September.

The IUCN lists the species as 'critically endangered'. Changes in agricultural practices in Europe and Asia have led to drastic changes of the habitat of the Sociablelapwing. In areas where grazing by big herds of sheep and cattle has been abandoned, vegetation grows dense and unsuitable for breeding. In other areas agriculture has intensified. The regular treatment of fields and the increased trampling of cattle and sheep destroy the nests. Birds of the corvid family, especially Rooks, have benefited from afforestation in Russia and Kazakhstan and predation on eggs and chicks of the Sociable lapwing has increased.

ABOVE:

MACCOA DUCK

Oxyura maccoa

The Maccoa Duck is a species confined to Africa. A northern population occurs in Eritrea, Ethiopia, Kenya, and Tanzania, and a southern population occurs in Angola, Botswana, Lesotho, Namibia, South Africa and Zimbabwe.

The global population estimate is 9,000 to 11,750 birds. The northern population appears to be in rapid decline, whereas the southern population is stable. Previously, the species was estimated to be more numerous and occurring in more countries, so the population size was overestimated. Movements are not very well known. Northern populations shift from the temporary wetlands occupied when breeding to deeper permanent waters when not breeding. Southern populations occur from sea-level to inland waters at high altitude. They probably move less than 500 km.

The IUCN lists this species as 'least concern'. Maccoa ducks are caught incidentally in gill-nets and drown. Rapidly changing water levels in impoundments or as a result of deforestation disrupt breeding and feeding conditions. Another threat is degradation of habitats by draining of wetlands, invasion by alien vegetation and pollution. As Maccoa ducks feed on invertebrates in the bottom sediment, the dose of pollutants that accumulate up the food chain might reach lethal or near-lethal levels. Disturbance and nest predation add more threats. The estimate of the population size and the rapid decline indicate that the species may soon be re-evaluated as 'near threatened' or 'vulnerable'.

BALD IBIS

Geronticus eremita

The range of the Bald ibis extends across Morocco and Syria, the Red Sea coast of Saudi Arabia, Yemen and Ethiopia, Eritrea and Djibouti. Its usual habitat is cliffs.

There are an estimated 420 left in the wild (but fewer than 100 breeding pairs) and a further 1,500 in captivity. Until recently it was thought that this species had been reduced to one non-migratory population in the Souss-Massa National Park near Agadir in Morocco after it had been declared extinct in Turkey in 1989 (it disappeared from the Alps 400 years ago). In 2002 a tiny migratory population was discovered in the Syrian Desert. It appears that these birds migrate south in winter down the west coast of Saudi Arabia and Yemen to Ethiopia, Eritrea and Djibouti in north-east Africa.

The IUCN lists this species as 'critically endangered'. A proposed tourist development near the Souss-Massa National Park, the species' one remaining stronghold in Morocco, could prove detrimental to the birds if not constructed in a sensitive way. Hunting and pesticide use may have contributed to the species' decline.

RUDDY-HEADED GOOSE

Chloephaga rubidiceps

The Ruddy-Headed goose lives in the South of Chile and Argentina and the Falkland Islands/Islas Malvinas. It is found in marshy wetlands (or "vegas") and open country, such as coastal grassland and meadows, commonly seen with Upland or Magellan geese (*Chloephaga picta*) and Ashy-Headed geese (*Chloephaga poliocephala*). Its diet consists of roots, leaves, stems and seed-heads of grasses and sedges.

Two populations of the Ruddy-Headed goose (*Choephaga rubidiceps*), which is the smallest austral goose inhabiting South America, have been identified. The sedentary population is confined to the Falkland Islands/Islas Malvinas while the mainland one migrates between its breeding grounds in Tierra del Fuego and southern Patagonia of Chile and Argentina and its wintering quarters in southern Buenos Aires province, Argentina. It is the latter population, which is in serious danger of extinction with an estimated size at around 900-1,000 individuals.

IUCN categorises the species as 'least concern'. Until as recently as the 1980s this species was considered a threat to agriculture and it has now been taken off the Argentine and Falkland Islands/Islas Malvinas pest species lists. Persecution in its wintering grounds was a major factor in the species' decline. The introduction of the Patagonian grey fox (*Dusicyon Griseus*) to control rabbit populations on Tierra del Fuego may also have had an impact on the Ruddy-headed goose. The larger and more aggressive Magellan or Upland goose also outcompetes the Ruddy-Headed goose where the two species coincide on foraging and nesting grounds.

WOOD SANDPIPER

Tringa glareola

With a global population estimated at between 3-3,5 million and habitat measured at over 15,000,0000 square kilometres, the Wood sandpiper is listed under the IUCN's lowest category of threat, 'least concern'. Nonetheless, the species is among the 255 included on the Annexes of the African-Eurasian Migratory Waterbird Agreement (AEWA), the largest instrument so far negotiated under CMS.

The Wood sandpiper is a solitary, monogamous creature, which inhabits peatlands, open swamps in northern forests, scrubland between the tundra and coniferous forests, wet heathlands and marshlands.

An adult Wood sandpiper stands approximately 20cm tall and has a wingspan of 35-40cm.

In winter, scattered groups of 50 congregate and flocks of 1,000 and more are known during migration. In late June, the adults start to leave the breeding grounds in the Scottish Highlands, Scandinavia, Northern Europe and Asia, with the juveniles following in August, arriving in the wintering grounds in tropical Africa and South Asia from late-July to October.

Threats to the Wood sandpiper include the degradation of its peatland habitats by excessive artificial drainage for agricultural and forestry purposes. Climate change may account for recent declines in the populations in Scandinavia, Finland and Germany. The species is also vulnerable to diseases such as avian botulism and malaria.

ABOVE:

GREAT BUSTARD

Otis tarda

The Great bustard is listed as 'vulnerable' by the IUCN and on both CMS Appendices. It is the heaviest bird capable of flight, the males weighing up to 16kg.

The species numbers between 35,600 and 38,500 individuals worldwide. Though there are other populations in Spain, Portugal, Morocco, Russia, Kazakhstan, China and Mongolia, the Great Bustard Memorandum of Understanding (MoU) covers the residual Middle-European populations.

Modern agricultural practice has caused a rapid decline in much of Central and Eastern Europe. Without the active protection measures outlined in the MoU's Action Plan, the species seems doomed to disappear from many of the Range States. The remaining population is dispersed in several small pockets. Its habitat is intensively-used agricultural land and mixed extensive agricultural and pasture or fallow land. Conservation measures need to focus on active habitat management and on maintaining large areas of non-intensive farming systems.

OPPOSITE:

HOUBARA BUSTARD

Chlamydotis undulata

There are two members of the *Chlamydotis* genus — the African Houbara (*Chlamydotis undulata*) and the Asian or Macqueen's Houbara Bustard (*Chlamydotis undulata macqueenii*). The geographical dividing line between the two is the Sinai peninsula in Egypt. These species breed in deserts and other very arid sandy areas. They are omnivorous, taking seeds, insects and other small creatures. Previously treated as two subspecies, they are now recognized as entirely separate species.

The Asian Houbara migrates south every winter from its breeding grounds in Central Asia, China and Mongolia. Its wintering habitats are found in Pakistan, Iraq, Iran and parts of the Arabian Peninsula, including the United Arab Emirates. Formerly, the entire Arabian Peninsula had a large resident breeding population, but this is almost extinct. Small breeding populations remain in Saudi Arabia, Yemen and Oman. There are approximately 50,000 Macqueen's Houbara bustards, most of which come from Kazakhstan and Uzbekistan.

Excessive hunting for falconry over the last few years is responsible for diminishing Houbara numbers. As a result, the IUCN lists the species as 'Near threatened', having been drastically reduced due to excessive exploitation to the point of becoming endangered, or even extinct, in some places.

AQUATIC WARBLER

Acrocephalus paludicola

This small waterbird has declined sharply at a rate equivalent to 40% in the last ten years. The Aquatic warbler is a regular but rare autumn migrant travelling up to 12,000 km from Eastern Europe to sub-Saharan Africa. Over half of the world population of this species breeds and spends part of the year in the marshes and fens of Belarus. Its dependence on specialised and vulnerable habitat means it has become globally threatened, as its habitats have suffered from constant decline. This decline is mainly due to human induced changes in the hydrological regime in key sites (both drainage and flooding), changes in land use and habitat fragmentation caused by infrastructure building. The effects of pollution pose a further threat. In 1992 population estimates indicated that there were 12,000-20,000 singing males.

CRANES
Gruidae

The Siberian crane (*Grus leucogeranus)* (below) is one of the three rarest crane species and probably the most endangered in the wild. The Siberian crane is found in Afghanistan, Azerbaijan, China, India, Iran, Kazakhstan, Mongolia, Pakistan, Russia, Turkmenistan and Uzbekistan. It breeds and winters in wetlands, preferring wide expanses of shallow fresh water with good visibility, where it feeds primarily on the shoots, roots and tubers of aquatic plants. The Siberian crane is the most aquatic of the crane species, as it nests, roosts and feeds in bogs, marshes and other wetlands of the taiga (forest) and tundra transitional zone.

The largest of the known flocks, accounting for well over 95% of the wild population, numbers 3,000 birds. It overwinters at Poyang Lake in China and breeds in eastern Siberia. The western flock numbers under 10 birds, breeds along the Ob river near the Ural mountains in western Siberia and spends the winter near the Iranian Caspian Sea shore. The last two birds in the central population that used to spend the winter at the Keoladeo Ghana National Park in Rajasthan, India, were last sighted in 2002. There have been occasional unconfirmed reports of sightings along the central population's flyway. Rare sightings of vagrants have been reported in Japan and Korea.

The traditional migratory and wintering habitats of this species are under constant pressure from the demands of the growing human population. These include: agricultural development, wetland drainage, oil exploration, hunting and water development projects. The western population is primarily threatened by hunting whereas the eastern population is at risk from the loss of its wetland habitat, and the species is listed as 'critically endangered'.

The Demoiselle cranes (*Anthropoides virgo*) pictured on the following pages, have also, like Siberian cranes, suffered from the loss of wetland habitat in key breeding areas, however, they are not threatened with extinction.

SHY ALBATROSS

Thalassarche cauta

Also known as the Tasmanian Shy albatross, White-capped albatross or Shy Mollyhawk, the Shy albatross breeds on three islands off Tasmania (Albatross Island, Pedra Branca and Mewstone). Most sightings of this species occur around Tasmania and Southern Australia but ringed birds (mainly juveniles from the Mewstone colony) have been seen as far away as South Africa. They are less pelagic than other albatross species and tend to stay within the continental shelf. Adults of breeding age tend to stay closer to the breeding sites. There is a breeding population of 13,000 pairs and a total population of perhaps 60,000 individuals.

All breeding sites are legally protected and the birds experience little human disturbance – the only contact tends to be with conservationists. Occasional viral outbreaks have affected chick mortality detrimentally. Severe wave action may affect albatrosses on Pedra Branca. Alien species and heavy metal contamination are not thought to be as serious a problem for this species as they are for others. Satellite tracking indicates that this species' range overlaps the areas covered by four regional seas fisheries agreements – notably the Convention on the Conservation of Southern Bluefin Tuna, Indian Ocean Tuna Commission and the Western and Central Pacific Fisheries Commission. As with all albatross, bycatch is the principal threat. Pollution, marine debris and ingestion of plastic are the other major factors.

ANDEAN FLAMINGO

Phoenicopterus andinus

The Andean flamingo is the rarest of six flamingo species and occurs on the high Andean plateaus of Peru, Chile, Bolivia and Argentina. It lives on alkaline and salt lakes at high altitude (2,300-4,000 metres). There are populations in ten known locations. As a possible result of El Niño, this species has just been recorded for the first time as breeding in Argentina. The population of 34,000 is decreasing year-on-year.

Eggs were taken as food from the middle of the twentieth century until the 1980s, drastically accelerating the birds' decline. Mining, adverse extreme water-levels (due both to heavy rainfall, drought and human intervention), erosion of nest sites and human disturbance have also been detrimental. Juvenile birds are occasionally targeted outside protected areas for food and their feathers. Eggs and chicks are also predated by the Culpeo fox (*Pseudalopex culpaeus*), particularly in dry periods when the nest sites are more accessible. As a long-lived slow breeding species raising usually no more than one chick per season, this species is slow to recover from cataclysmic events.

LESSER KESTREL
Falco naumanni

The Lesser kestrel is found across the Middle East, from Morocco to Central Asia, and in China. It has disappeared from the Ural regions of Russia and Kazakhstan. It often nests in large colonies close to human settlements in building niches, on cliffs or in holes in trees. It feeds in habitats like steppes, natural and managed grasslands, and non-intensive agricultural lands. There are 50,000-60,000 in the South African population and 20,000 breeding pairs in the Mediterranean and North African population. It is declining, having experienced a substantial fall (possibly 50%) since the 1970s. The northern population winters in Southern Spain, Southern Turkey and Malta and across North Africa.

The Lesser kestrel is listed as 'vulnerable' in the IUCN Red Data list. The main cause of its decline in Europe is habitat loss and degradation in its breeding grounds, mainly resulting from intensive agriculture, urban growth and afforestation. A similar picture occurs in South Africa, where its grassland habitat is losing out to agriculture, afforestation and intensively farmed pasture. Pesticides may directly poison the birds, but they may also be taking a heavy toll on the falcons' insect prey. The restoration of old buildings can reduce the availability of suitable nesting sites.

EURASIAN BLACK VULTURE

Aegypius monachus

The Eurasian (or European or Cinereous) vulture is found across the Iberian Peninsula, the Balkans, Turkey, the Caucasus, Iran, Northern India, Siberia, Mongolia and China. It winters in Sudan, the Middle East and Korea. In Europe, adult birds are non-migratory, in Central Asia they follow nomads and in Eastern Asia, the birds are partly migratory.

The main food is carrion from medium- to large-sized carcasses. Only very rarely do Black vultures take live prey such as lizards and tortoises. In Spain, it favours rabbits and sheep, in Mongolia it thrives in areas where the Bobac marmot *Marmota bobak siberica* is affected by the disease Tarbagan pest, and in the Himalayas, it feeds on yak, sheep and gazelles. Its strong bill enables it to rip through muscles, tendons and skin.

In the breeding season the Black vulture congregates in loose colonies with occupied nests being no closer than 30 metres apart (sometimes as far as 2 km). It usually nests 1.5-5 metres high in trees – preferring oaks in Spain and Portugal, junipers in the Caucasus or almond trees in Uzbekistan. Very rarely in Europe, but more often in Asia, it nests on rocks. The nests are large constructions made of sticks often nearly 2 metres wide and between 1 metre and 3 metres deep. Usually each clutch is made up of one egg, sometimes two.

The Eurasian Black vulture appears on Annex I of the CMS Memorandum of Understanding on the Conservation of Migratory Birds of Prey in Africa and Eurasia signed in Abu Dhabi on 22nd October 2008 by twenty-eight Range States. The species is considered 'vulnerable' and is listed on Appendix II of CITES. There is a trade in vulture feathers in China and taking live specimens is a problem in parts of Central Asia.

The main threats are the destruction of the forests where it breeds, lack of food and direct persecution. It has disappeared from much of its former range in the West, except for Spain where numbers rose from just 200 pairs in the 1970s to approaching 1,000 pairs in the early 1990s. There are remnant populations in the Balkans and fewer than 100 pairs in the countries of the Caucasus.

BONELLI'S EAGLE

Hieraaetus fasciatus (or sometimes Aquila fasciata)

Bonelli's eagle is found in scattered areas across Iberia, North Africa, the Balkans, Middle East, India, Pakistan and Afghanistan, China and Indo-China and Indonesia. It grows to a height of 72 cm with a wingspan of up to 180 cm and weighs between 1.5 and 2.5 kg. Females are larger than males and the species is noted for the white patch on its back.

The species' preferred habitat is mountainous or broken terrain in sunny, warm climates. Vegetation cover varies considerably from barren slopes to areas with many bushes and shrubs. It is normally found in lower altitudes but is seen at 2,000 metres in Africa and Asia. It seems well capable of adapting to human presence.

Bonelli's eagles have a varied diet. In Europe they prefer rabbits and game birds but when these are scarce they will take crows, squirrels and reptiles instead. Agile fliers, they tend to take prey on the ground but sometimes they catch birds in flight. They sometimes hunt in pairs.

The egg-laying season is February to March in the Mediterranean but earlier in India. They build large bulky nests usually on rocky cliffs but sometimes also in trees. Clutches of one or two (and rarely three) eggs are incubated for five to six weeks by the female. Chicks fledge after nine to ten weeks and juveniles may establish themselves hundreds of kilometres away from where they were born. Pairs tend to stay close to their home range during the breeding season, venturing further afield at other times.

Bonelli's eagle is listed on Appendix II of CITES and is not considered to be globally threatened. All migratory members of the *Accipitridae* family (which includes hawks, eagles, kites, harriers and Old World vultures) are included on Appendix II of CMS. Bonelli's eagle is not however included on the Annex to the CMS MoU on Birds of Prey in Africa and Eurasia. It is in decline across Europe, and in the 1980s, while populations of other eagle species recovered in Spain, Bonelli's eagle numbers dropped. Direct persecution and accidents involving power-lines are major problems as are habitat degradation and reductions in the numbers of prey species.

EURASIAN EAGLE-OWL

Bubo bubo

The Eurasian Eagle-owl, also known as the Common, Great or Northern Eagle-owl is noted for its prominent ear-tufts, powerful beak and feet. It can grow to 75 cm with a wingspan of 190 cm. Females which can reach 4kg are larger than males which normally weigh 1.5-2.8 kg.

The Eagle-owl prefers rocky terrain with cliffs, ravines and woodlands, such as the Russian taiga, as a habitat and it forages over flood plains and other open, sparsely wooded areas such as farmed valleys. It is found in large parts of Europe and across Central Asia. The species is particularly sensitive to disturbance and adults will abandon nests and chicks.

Its prey is varied, comprising small mammals like the water vole and hares, birds such as jays, herons and even smaller raptors and, more rarely, reptiles, insects and amphibians. Preferred prey ranges in size-from 200 to 2,000 kg. The diet varies with the seasonal abundance of prey species. Cannibalism is not unknown, where the weakest chick is eaten by its siblings or parents.

The Eagle-owl hunts mainly in the evening or at night, but at the northern edge of its range it is often active during the day. It normally hunts from a perch but also scouts for prey when flying close to the ground. It attacks its victims by surprise.

In most of its range states the Eagle-owl is resident. Juveniles disperse over considerable distances, with vagrants recorded as far away as the Nile. At the northern extreme of the range, there is some seasonal movement south in winter to find food sources and to avoid the harshest weather conditions. Population densities in most areas are very low, with just a few pairs in every 100 square kilometres being typical.

The Eagle-owl is not considered to be globally threatened and is on Appendix II of CITES. Populations suffered a rapid decline in the 20th century due to direct persecution and as a result of collapsing rabbit populations after outbreaks of myxomatosis. Road accidents and collisions with barbed-wire fences also took their toll. Strict legal protection and reintroduction programmes saw the species recover in the 1970s but overall there are still considerable downward pressures and despite the protected status the Eagle-owl is still persecuted in some localities.

2

AQUATIC SPECIES

The conservation of marine mammals listed by the CMS is a great challenge, in particular because these species are affected by multiple threats, often within international waters. CMS has adopted a regional approach with promising results. Four CMS agreements are engaged in different areas of cetacean conserv-ation, while further agreements exist for the conservation of the monk seal and dugong.

The conservation of marine mammals listed by the CMS is a great challenge, in particular because these species are affected by multiple threats, often within international waters. CMS has adopted a regional approach with promising results. Four CMS agreements are engaged in different areas of cetacean conservation, while further agreements exist for the conservation of the monk seal and dugong.

The Agreement on the Conservation of Small Cetaceans of the Baltic, North East Atlantic, Irish and North Seas (ASCOBANS) aims to conserve small whales, dolphins and porpoises such as the once-familiar Harbour porpoise and the spectacular Orca. The most important threats facing these species of toothed whales are incidental capture in fisheries, collisions with ships, acoustic disturbance and marine pollution. Under the auspices of ASCOBANS, the Jastarnia Plan, the Recovery Plan for Harbour porpoises in the Baltic Sea, is the result of a collaborative effort of a series of scientific initiatives and meetings over several years. The main focus of this recovery plan is the identification of human induced threats to the recovery of the species.

The second Agreement on whales and dolphins developed under the Convention is the Agreement on the Conservation of Cetaceans of the Black Sea, Mediterranean Sea and Contiguous Atlantic Area (ACCOBAMS). Its main aim is to reduce threats to small and great whales, such as the Fin and the Sperm whale. Conservation Plans under this agreement provide, among other things, for the assessment of human-cetacean interactions, emergency response measures, the establishment of protected areas and the reduction of negative interaction with fisheries.

Migration routes of marine mammals pass through the coastal waters of countries as well as the high seas. The relevant Memorandum of Understanding for the Conservation of Cetaceans and their Habitats in the Pacific Islands Region covers all populations of whales and dolphins in the Pacific Islands region, which have not yet recovered to pre-whaling levels. This framework of CMS helps the countries to standardize the conservation and the educational programmes for local communities and commercial fishing.

A number of species of small cetaceans can be found in West African waters, including the endemic Atlantic humpback dolphin. Small cetaceans, which include dolphins, porpoises and small toothed whales, are subject to various threats, such as habitat degradation, bycatch, direct catch, over-fishing and pollution. In order to study and provide information on the conservation status of small cetaceans in West Africa, CMS started an initiative for the Conservation of Marine Mammals in Western Africa. Western African Talks on Cetaceans and their Habitats (WATCH) are a series of scientific and inter-governmental meetings on marine mammals. The aim is to develop an Action Plan for the conservation of West African small cetaceans and manatees under the Memorandum of Understanding of West African Small Cetaceans and Sirenians of the Eastern Atlantic Basin.

The Agreement on the Conservation of Seals in the Wadden Sea was concluded as the first Agreement under CMS after an epidemic in 1988 killed 60% of the region's Harbour Seals. The Agreement has proved successful: the population has regained its pre-epidemic levels and, although still subject to diseases, the seals are no longer threatened with extinction. The aim is to restore and maintain viable stocks, and increase reproductive capacity, including improved survival rates among juvenile seals.

The Mediterranean Monk seal is one of the most threatened marine mammals in the world. The Memorandum of Understanding concerning conservation measures for the Eastern Population of the Monk seal aims to save the last few animals (approximately 500) remaining in the wild. Recovery of the depleted population and reducing habitat loss are the main focus.

The dugong is a large strictly marine, herbivorous mammal. The Memorandum of Understanding covering the Indian Ocean is designed to conserve the populations and their habitats from detrimental anthropogenic influences like hunting or agricultural and industrial runoff into the waters they live in.

Marine turtles are among the oldest vertebrate life forms on Earth. They are threatened by by-catch, unsustainable consumption of both meat and eggs, degradation of the coastal environment, climate change and marine pollution. Little is known about their lives in the open ocean. They provide a perfect example of the need to bring together local communities, conservationists, researchers and government authorities to work in a co-ordinated way. There are two CMS Memoranda of Understanding concerning marine turtles: one for the Atlantic Coast of Africa and another for the Indian Ocean and South-East Asia (IOSEA).

Fish in troubled waters

The European sturgeon is an anadromous migrant, meaning the adults leave the sea to swim up rivers to reproduce. They are sensitive to any physical barriers to their migration and are severely affected by physical and chemical changes to watercourses. The Action Plan developed strategies to assist the wild population with captive breeding and release schemes. The Whale shark is the largest living fish species, growing up to 14 metres long. It inhabits the open sea in tropical and warmer temperate waters and feeds on plankton. Seasonal feeding aggregations of the sharks occur at several coastal sites. Though it is usually seen out at sea, it has also been found closer to shore, entering lagoons or coral atolls, and near the estuaries of rivers. CMS is leading efforts to develop a global conservation instrument for migratory sharks. With the help of a Memorandum of Understanding, threats like illegal trade and fisheries by-catch could be reduced. This initiative is particularly urgent given that the annual take of sharks worldwide has been estimated to exceed 100 million individuals.

Previous Page:
A mother and her calf, West Indian manatee, (*Trichetus manatus*) migrate along the Crystal River, Florida, USA.

Opposite:
The walrus (*Odobenus rosmarus*) is immediately recognisable by its huge tusks. There are three subspecies of walrus, each of which migrates from warmer latitudes to cooler each spring.

HARBOUR PORPOISE
Phocoena phocoena

Harbour porpoises are found in temperate and sub-polar coastal waters of the northern hemisphere. While some populations are resident, significant seasonal movements occur, prompted mainly by the availability of food. Migratory porpoises tend to spend the summer inshore and move further offshore in winter. Others however spend the summer in the north of their range and the winter in the south. Estimates of the European population vary widely, between 20,000 and just under 100,000.

The population has been substantially reduced by human activities in and near the porpoises' coastal habitat, most notably through bycatch. They are outcompeted for food by human fisheries and are susceptible to chemical poisoning. Studies in the Bay of Fundy in eastern Canada revealed high levels of cadmium, lead, mercury, PCB and DTT in the blubber, muscles and internal organs of specimens examined. The species is also prone to strandings.

The Harbour porpoise is on CMS Appendix II and is one of the key species – along with the Bottle-nosed dolphin (*Tursiops truncatus*) – covered by ASCOBANS (Baltic, Irish and North Seas and the North East Atlantic). It is also found in the Mediterranean and Black Seas and is therefore protected under ACCOBAMS. Different sub-species occur in the Atlantic, the Pacific and the Black Sea and its offshoots. Research is still being carried out to verify whether the Baltic Sea population is also distinct from neighbouring ones.

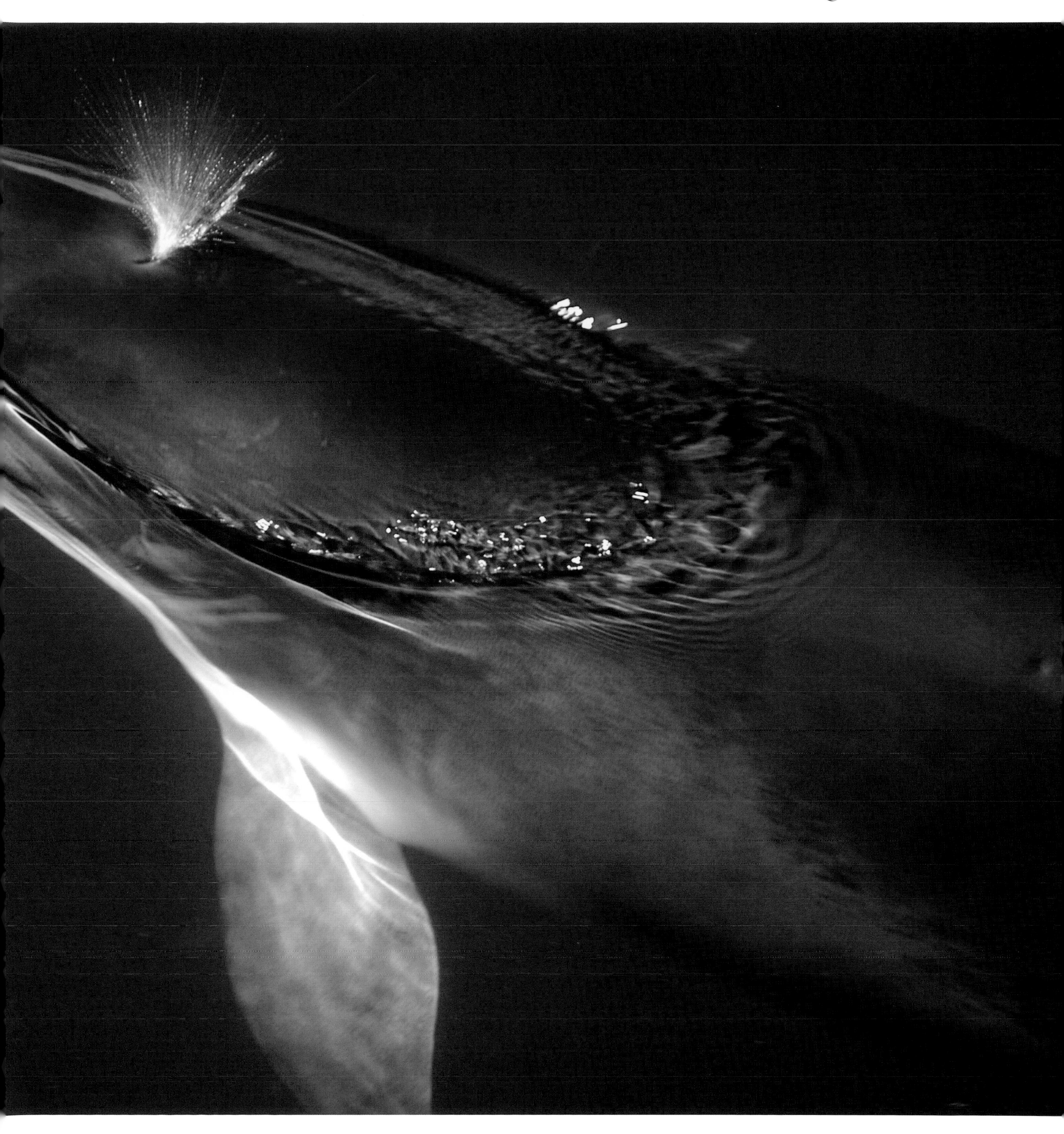

COMMON (OR HARBOUR) SEAL

Phoca vitulina vitulina

Occurs widely on the shores of Denmark, Germany, The Netherlands and the British Isles. The trilateral conservation area covered by the Agreement includes, in The Netherlands, the areas under Key Planning Decision Wadden Sea, in Germany, the Wadden Sea national parks and the protected areas under the Nature Conservation Act seaward of the main dike and the brackish water limit including the Dollard, in Denmark, the Wildlife and Nature Reserve Wadden Sea.

Disease – there were severe outbreaks in 1988 and 2002 of phocine distemper, the first of which claimed the lives of 18,000 seals in Northern Europe alone with the second claiming over 20,000 in the North Sea (51% of the estimated population); disturbance through human activities and marine pollution are among the other main concerns.

The Agreement has adopted a Work Plan which covers the Wadden Sea stock of the Common seal (*Phoca vitulina vitulina*) and is also extended to cover the two breeding stocks of the Grey seal (*Halichoerus grypus*) in the Wadden Sea, the latter not covered by the Wadden Sea Seal Agreement. The overall aim is to restore and maintain viable stocks and a natural reproduction capacity, including improved survival rates among juvenile Common and Grey seals. The Plan specifies actions in the following areas: conservation and management measures regarding habitats, pollution and wardening; research and monitoring; taking and exemptions of taking; and public information.

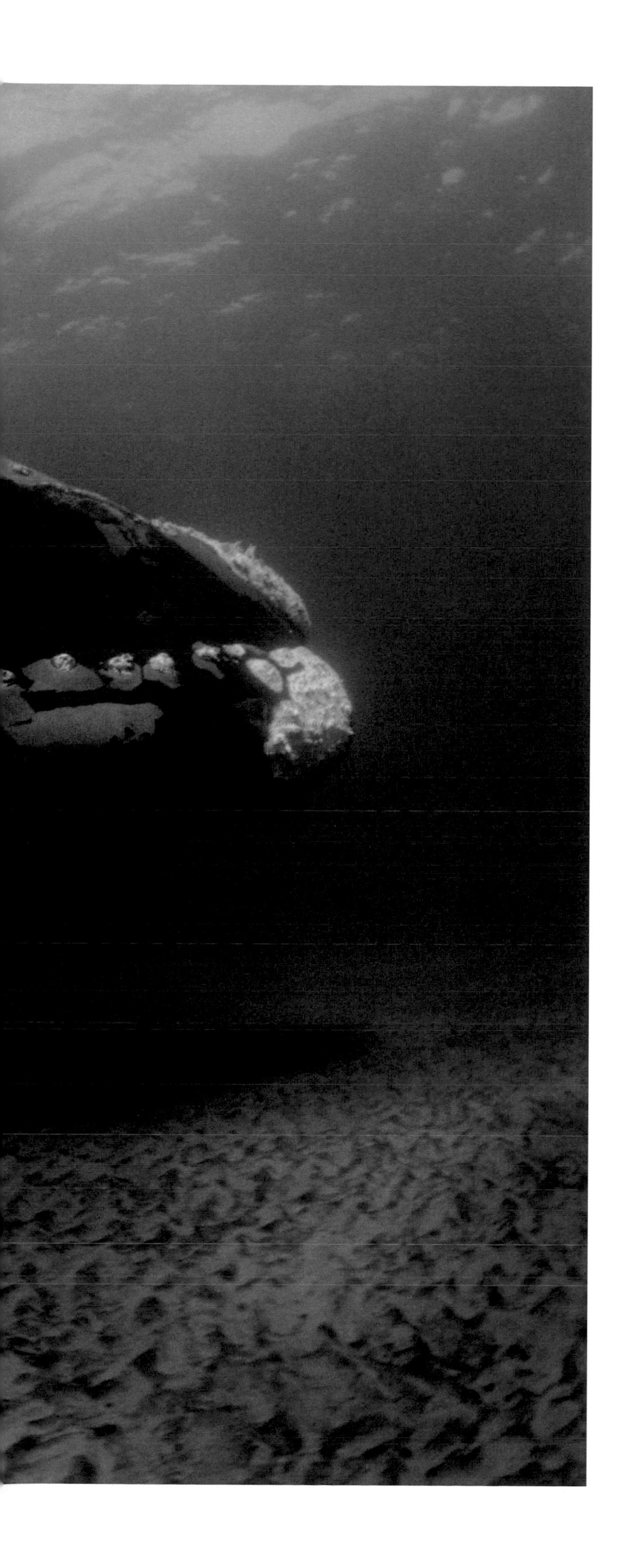

RIGHT WHALE
Eubalaena

Right whales were so named because they were the most lucrative whales to hunt as they provided large quantities of oil. They eat plankton and krill, scooping them up with their mouths wide open and filtering out the water through their baleen (or whalebone). Adult Right whales grow to between 11 and 18 metres (35-55 feet). There are three species of Right whale — the North Pacific (E*ubalaena japonica*), the Northern (*Eubalena glacialis*) and the Southern (*Eubalaena australis*).

Right whales frequent the North East Atlantic around the British Isles, Norway and Iceland and the east coast of Canada and the USA. 300-350 individuals are thought to exist. They prefer colder, pelagic waters for feeding and move to warmer coastal waters for breeding and calving. It is not entirely clear where females go in years when they are not calving. The Right whale is very slow moving (its normal top speed is just 7 kmph) but it can be very acrobatic, leaping from the water and hitting the surface heavily with its stomach during the mating season. This behaviour may also be associated with attempts to get rid of parasites. This species is particularly vocal during the breeding season and uses a wide number of low frequency noises to attract a mate.

North Atlantic Right whales are listed as 'endangered' in the IUCN Red Data book despite hunting being totally prohibited since 1937. Hunting of the species is thought to have begun as long ago as the 10th century by Basque whalers operating in the Bay of Biscay. While they are no longer hunted, they still face conservation problems including collisions with shipping, conflicts with fishing activities, the destruction of their habitats through activities such as oil drilling, and even competition for food from other species of whales. Their reduced numbers and slow reproductive rate (the females can only produce a calf once every three years) also place the survival of this species in the balance.

A global ban on hunting Right whales was first imposed in 1937 but it was not strictly adhered to, with a number of whaling nations continuing to take Right whales as late as the 1960s. To counteract the alarming number of deaths caused by collisions with shipping, the authorities of the United States of America have imposed a compulsory reporting system for all vessels visiting US ports. Speed limits in certain channels during sensitive parts of the year (e.g. just after calving) have been considered. The more numerous Southern Right whale is the focus of the booming whale watching tourist industry with important centres in Hermanus (South Africa), Imbituba (Brazil) and the Peninsula Valdes (Argentina).

BOTTLE-NOSED DOLPHIN

Tursiops truncatus

Bottle-nosed dolphins are found in most temperate and tropical waters; they are found on the high sea, in coastal waters with heavy surf, estuaries and shallow lagoons. They normally live in small groups of up to 15, although occasionally they are found in larger schools of up to 600.

High sea populations migrate in spring to coastal waters and in the autumn back out to the open sea. The worldwide population is estimated to be in excess of 100,000.

The Bottle-nosed dolphin is listed as 'Data Deficient' in the IUCN Red Data List. There are some areas in the world where Bottle-nosed dolphins are directly targeted (e.g. by fishermen who are thought to kill the species because of perceived competition for fish and the damage the dolphins do to fishing gear). Bycatch is a world-wide problem – the use of gill and purse-seine nets off Peru is thought to account for the deaths of 100 individuals per year. Foreign gillnet fisheries off Australia, tuna fisheries, shark fisheries off South Africa, trawling in the Western Mediterranean and unidentified fisheries in the English Channel all take their toll. Interactions with fisheries are also thought to contribute to strandings. Tests show that dolphins in the Western Mediterranean have been contaminated with DDT and other poisons, many of which affect male fertility. High levels of mercury were found in stranded specimens on the coast of Corsica. Other poisons such as tributyltin (TBT) and butyltin (BT) were found in dolphins resident in Asian waters.

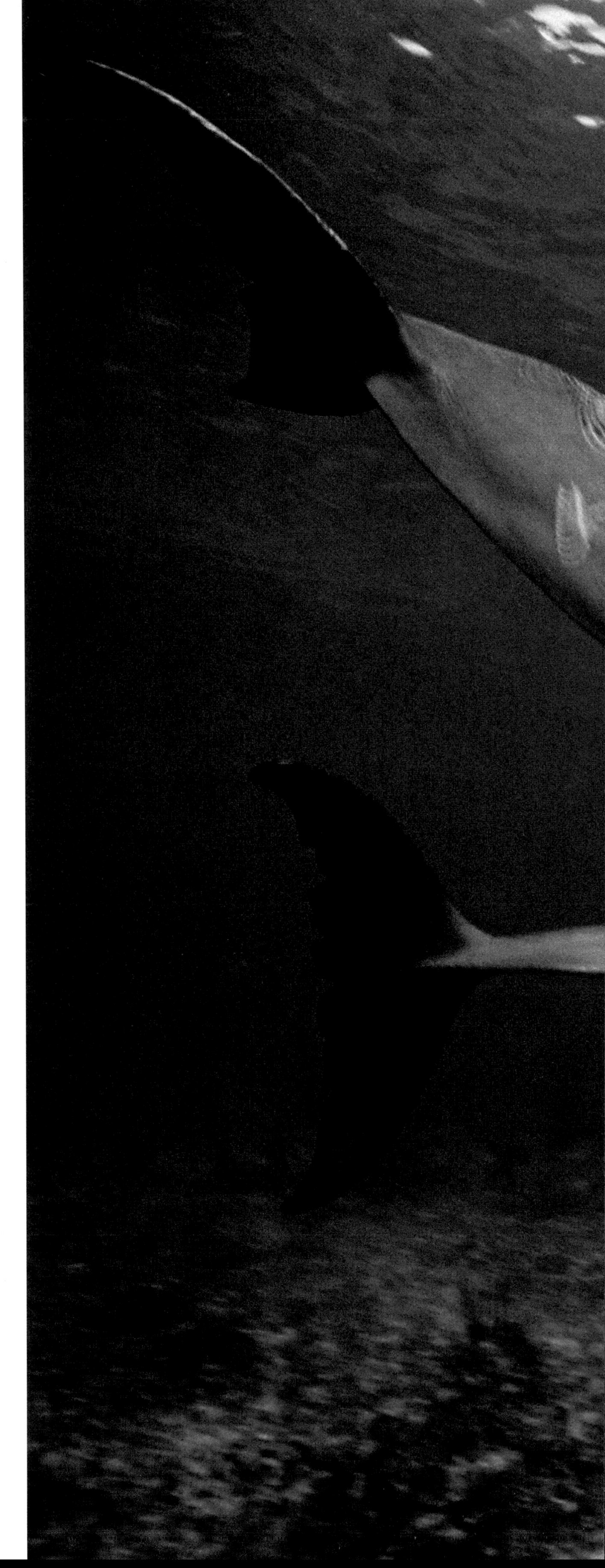

MONK SEAL

Monachus monachus

The three species of Monk seal are the Mediterranean, the Hawaiian (pictured opposite) and the Caribbean. The Mediterranean Monk seal is 'critically endangered', the Hawaiian is endangered and the Caribbean was declared extinct in 1996.

Once found throughout the Mediterranean and the Black Sea, the north-west coast of Africa, the Azores, the Canaries, Madeira and Cape Verde, the Mediterranean Monk seal is now restricted to a few pockets along the African coast, the Canaries and Madeira and in the Eastern Mediterranean around Greece and Western Turkey.

The Memorandum of Understanding (MoU) covers the Eastern Atlantic populations of the Mediterranean Monk seal. As one of the most threatened species of migratory marine mammal, the Monk seal is listed on both appendices of CMS. It is estimated that only 500 specimens remain in the wild. The species has disappeared from most of its historic range and only two breeding groups remain in the Eastern Atlantic — one on Madeira and the other on the Cabo Blanco peninsula. IUCN has classified the species as "critically endangered". The main threats to the seals are entanglement and mortality in fishing gear, overfishing and direct persecution from humans as well as natural factors like toxic phytoplankton and the deterioration of breeding sites (such as the roofs of caves collapsing). The isolation of the small remaining populations — experts believe that the two colonies do not interbreed — means that the species' survival is precariously balanced.

Since 1986 the different populations of the Mediterranean Monk seal have been the centre of the Mediterranean Action Plan of UNEP and CMS's own marine mammal activities. The Barcelona Convention and CMS will continue to liaise and coordinate their activities.

DUGONG

Dugong dugon

The dugong is one of four remaining members of the Sirenian family, along with three species of manatee. Unlike the manatees which also frequent fresh water, the dugong is confined to marine habitats, but like the manatees it is strictly herbivorous. Dugongs are particularly vulnerable to man-made influences, and due to their life-cycle (they are late and slow breeders), their extensive range and their distribution along rapidly changing coastal habitats. The seagrass pastures where they graze are threatened by eutrophication caused by agricultural and industrial run-off. Dugongs are also hunted for their meat and blubber and they are frequently struck by motorised vessels leading to injury or death.

The scattered nature of the different dugong populations means that the species is threatened with local extinction in many parts of its range. Overall, the IUCN classifies the species as 'vulnerable'. The dugong's closest relative, Steller's sea cow, was hunted to extinction in the 18th century. Australia hosts the largest populations of dugong, but they are also found in smaller numbers across the Indian Ocean and Western Pacific.

Protecting the vulnerable coastal habitats which support the sea grass pasture on which the dugongs feed will be an important element of concerted actions. Dugongs are normally found in wide, shallow, protected areas such as bays, mangrove channels and the lee sides of large inshore islands.

WHALE SHARK

Rhincodon typus

The Whale shark is the largest living fish species (growing up to 14 metres long), and the only member of the *genus Rhincodon*. It has existed in more or less its current form for 60 million years. It lives in the open sea in tropical and warmer temperate waters. While largely pelagic, the Whale shark does periodically congregate for feeding — such gatherings are known to take place off Western Australia, the Philippines, Zanzibar and Honduras.

There are no reliable population estimates, but there is evidence to suggest that Whale sharks are highly migratory with their movements determined by the occurrence of food supplies and the presence of certain geographic features. One individual that was tracked was found to have travelled over 13,000 km over the course of three years, this despite the fact that the Whale shark is not considered to be a particularly efficient swimmer.

EUROPEAN STURGEON

Acipenser sturio

Once found along the coasts and in the coastal rivers of most of Europe, its range is now restricted to the North Sea, the United Kingdom and the Atlantic coast of France. Only one breeding population remains, the one living in the basin of the Gironde, Dordogne and Garonne rivers. The largest migratory fresh water fish in Europe, it is a long-lived species and reaches reproductive maturity relatively late (10 years for males and 13 for females).

There are thought to be approximately 1,000 individuals spread across the remaining range. The European sturgeon is an anadromous migrant, meaning the adults leave the sea to swim up rivers to reproduce.

The European sturgeon is amongst the most threatened fish in Europe, as it is in critical danger of extinction. The population is fragile and genetically limited. Physical and chemical changes to watercourses caused by gravel extraction or pollutants can adversely affect the sturgeon. They are especially sensitive to any physical barriers to their migration. Competition for food from other species, especially other species of sturgeon, and poaching are also significant factors.

MARINE TURTLES

Marine turtles face threats from unsustainable exploitation, destruction of nesting and feeding habitats and incidental mortality in fishing operations. Some key nesting beaches are threatened by development. Endangered across the world, marine turtles are considered flagship species for activities aimed at protecting habitats important to other forms of marine life.

The conservation and management of marine turtles globally and within the Indian Ocean-South-East Asian region pose a formidable challenge for sustainable development objectives. Many communities still eat marine turtle meat and eggs and use the shells for traditional crafts. At the same time, marine turtles have both intrinsic and ecological value as components of marine ecosystems.

The species covered by the MoU are: the Loggerhead turtle (*Caretta caretta*), Ridley turtle (*Lepidochelys olivacea*), Green turtle (*Chelonia mydas*), Hawksbill turtle (*Eretmochelys imbricata*), Leatherback turtle (*Dermochelys coriacea*) and Flatback turtle (*Natator depressus*).

GREEN TURTLE
Chelonia mydas

The Green turtle (pictured left) has a wide distribution in subtropical and tropical seas across the entire world. It is also known as the White or Black turtle and its shell can be olive green or black. It owes its most frequently used common name to the green fat found under the shell.

It is the only herbivorous marine turtle with a diet consisting mainly of sea grasses and algae, playing a crucial role in maintaining the balance of the ecosystem. Green turtles can reach weights of up to 230 kg.

Sometimes also known as the "edible turtle" or "soup turtle" in other languages, this species has been eaten for centuries in Green turtle soup. Oil has also been produced from body fat for use in cooking, cosmetics and medicine. Partly as a result of harvesting of both turtles and eggs, the Green turtle is now considered 'endangered' on a global scale. About 150 separate nesting areas scattered around the world are known, with fewer than 20 hosting more than 2,000 nesting females. There are no estimates of total population size. Fortunately some populations, for example those in Florida and Malaysia, are showing signs of recovery as a result of conservation action.

The main threats facing the Green turtle are direct taking for their meat and eggs, which are considered a delicacy in some countries. Pollution is also a problem affecting both individual animals and populations as a whole. As with sea birds and cetaceans (whales and dolphins), bycatch in fishermen's nets set for other target species takes a toll with entangled turtles unable to escape and drowning. Construction along the sea shore, e.g. for tourist centres, leads to the destruction of nesting beaches as well as higher rates of mortality among hatchlings which, confused by street lighting, fail to find their way to the sea and fall victim to predators.

The Green turtle is protected under two CMS Memoranda of Understanding covering the Indian Ocean and South-East Asia and the Atlantic Coast of Africa. It is also listed under the Inter-American Convention for the Protection and Conservation of Sea turtles and CITES.

LOGGERHEAD TURTLE

Caretta caretta

The Loggerhead is a large species of turtle with a red-brown carapace or shell. Its head is much larger in relation to body size compared to other marine turtles, and its mouth accommodates powerful jaws.

Loggerhead turtles are found across the world in the waters of warm temperate and tropical zones. They are the only marine turtles to nest extensively in temperate areas and they are found regularly around the coastlines of Europe. Loggerheads are the most common turtle in the Mediterranean. The two largest nesting areas for the Loggerhead are in Oman, mainly concentrated on Masirah Island where each year up to 30,000 females come ashore to nest between April and July, and along the coast of the USA, on the shores of states along the coast of the Atlantic and Gulf of Mexico from Texas to North Carolina, with the largest concentration in Florida. Beaches in Greece, Turkey and Cyprus in the eastern Mediterranean support the third largest nesting population. In the case of Loggerheads, unlike with other turtles, mating takes place along their migration route rather than near the nesting beaches. The females lay clutches of around 100 eggs the size of table-tennis balls. The hatchlings emerge after sixty days, usually at night to reduce the risk of predation by sea birds.

Caretta caretta is categorised as 'endangered' on the IUCN Red Data List, and is included on Appendix I of CITES and on both Appendices of CMS. Both of the Memoranda of Understanding concluded under CMS for marine turtles – covering the Indian Ocean and South-East Asia in one case and the Atlantic coast of Africa in the other – seek to conserve the Loggerhead. The major threats faced by this species include the taking of eggs, bycatch, the accidental capture by commercial fishing boats and the degradation of nesting beaches, sometimes as a result of tourist developments. The bright lights of buildings along the shoreline can disorientate hatchlings. It is not unusual for individuals surviving until adulthood to reach the age of thirty. In exceptional cases ages well in excess of one hundred have been recorded.

Loggerheads have a varied diet eating bottom-dwelling molluscs, urchins, crabs and jellyfish. They are apparently unaffected by the poisons of Portuguese Men o' War as they have been known to feed on them too. Unfortunately young turtles suffer from the effects of eating plastic bags, which they can mistake for jellyfish, tar and other dangerous forms of discarded waste. These either poison them or become stuck in their gut.

KILLER WHALE (OR ORCA)

Orcincus orca

Orcas are the largest of the dolphins, with males growing to 9 metres and weighing 5,500 kg. They are easily recognisable by their distinctive black, white and grey markings including a white patch just behind each eye.

Orcas are found in virtually all of the world's oceans from the Arctic and Antarctic through temperate zones and around the Equator. They are however most numerous in coastal waters, with most sightings within 800km of the shore, and in cooler climes. They prefer deep water but are occasionally found in shallow bays or even in rivers. There is some evidence to suggest that certain populations are sufficiently different to warrant classification as subspecies.

There are recognised populations off the west coasts of Canada, the USA and Mexico; Iceland; Norway; Japan; the Antarctic and the Southern Indian Ocean ranging from a few hundred in the North East Pacific to 80,000 in the far south. Migration behaviour varies considerably with some pods undertaking long movements while others simply move in search of prey. They can travel as far as 200 km a day while foraging.

Social organisation is based on pods to which the orcas remain loyal throughout their lives. Pods can be made up of as many as 50 animals. Non-migratory groups tend to be larger than the transient ones. Members of the pod show complex cooperation while hunting. Target species vary considerably, with each pod developing its own specialisms, including other marine mammals such as seals, various fish species and cephalopods (squid). It has been recorded that orcas followed fishing vessels for over a month across 1,000 miles feeding off discarded fish.

The threats to the species used to include low level direct take by whalers until the early 1980s. Live capture for export by Iceland was discontinued in 1991. Conflicts with commercial fisheries have resulted in orcas being shot. Bycatch in fishing gear does occur but is rare. Chemical pollution appears to be a problem with high levels of PCBs and DDT found in the blubber of specimens off Washington State. Orcas have been seen with propeller scars and ship movements and motor noise can disturb all cetacean species by interfering with echolocation signals. Overfishing and local disappearance of prey may have resulted in the decline of some populations, although orcas are opportunistic and are not dependent on any one species for food.

The Killer whale is included on Appendix II of CMS. It is one of the species covered by the CMS instruments, ASCOBANS (small cetaceans of the North Sea, Irish Sea, North-East Atlantic and Baltic), ACCOBAMS (Mediterranean and Black Sea) and the Memoranda for West African Marine Mammals and the Pacific Islands region cetaceans.

WALRUS
Odobenus rosmarus

The walrus is the only member of the genus *Odobenus*, part of the pinniped family of fin-footed animals, which also includes seals and Sea lions. There are three subspecies of walrus – the Atlantic, the Laptev and the Pacific – confined to the neo-Arctic regions around Greenland-Canada, Northern Russia, and Alaska-Northeastern Russia respectively.

In the breeding season, walruses frequent ice floes. At other times, they form large colonies on rocky beaches and outcrops. Walruses are noted for their large tusks – they are the only pinniped species to have them, their great size – males can reach a weight of 2,000 kg and even newborn pups weigh around 100kg – and their moustache-like whiskers.

The walrus is an opportunistic feeder, which preys on shrimps, crabs, clams, sea cucumbers and molluscs. They can dive to depths of up to 80 metres and stay under water for thirty minutes.

The name "walrus" is of Germanic derivation and means "whale horse". The scientific name is a compound of odous (tooth) and baino (walk) and was coined because walruses appear to drag themselves out of the water using their tusks. The tusks are also used for keeping air holes in the ice open and for fighting, not uncommon during the rutting season.

Indigenous peoples have hunted the walrus for their meat, bones, skin and blubber for centuries. Commercial hunting in the 19th and 20th centuries caused the population to plummet. Numbers have recovered, although populations are still fragmented. The Pacific subspecies is the most numerous with 200,000 individuals. The Atlantic population was nearly eradicated by excessive hunting and is perhaps a tenth this size and the isolated Laptev population is estimated at just 5,000 to 10,000.

Like the Polar bear, the walrus is facing the depletion of its favoured habitat due to global warming and climate change. The impact of the reduction of the ice sheets where females give birth and nurse their pups has yet to be quantified.

Under CITES, the trade in walrus ivory is restricted through the species' listing on Appendix III. The walrus is not listed on the Appendices of CMS. IUCN lists two of the subspecies as 'least concern' and the third as 'data deficient'. Russian populations are listed as 'decreasing' and 'rare' in the national Red List.

3

TERRESTRIAL SPECIES

The great land migrations speak to the romantic in us all: vast numbers of elephants, wildebeest, zebra crossing seemingly endless plains to reach, eventually (and not for all) water and food and a place to rest. These animals, often the larger species, arguably come into the greatest conflict with man, traversing his invisible borders and travelling into the hands of the poachers who lurk in less protected areas.

The less well known species, including many bats and butterflies, face similar threats. The following pages offer a survey of some of the more endangered species, and outline some of the measures that can help reduce the threats.

Numerous different migratory land mammals from apes to zebras regularly cross national borders. Several are endangered and therefore listed on the Convention's Appendices, such as bats, the magnificent Snow leopard, the Bactrian camel, the Gorillas, African elephants, deer species, and several antelope species in Africa and the Saiga antelope in Eurasia. Just as these animals vary widely, so do the strategies for their conservation.

Major conservation initiatives

Gorillas build new nests each day at dusk and move on at dawn to new areas of forest. Despite their relatively small population sizes compared to other migratory species covered by CMS, the ranges of gorilla populations frequently cover several countries. Destruction or modification of their habitat by deforestation, woodland exploitation, increasing demand for arable land and also the development of infrastructure such as forest roads are the main threats. Unstable political climates, armed conflicts, viral epidemics, illegal killing for trophies and for bushmeat, kidnapping infant gorillas for zoos and the exotic pet trade, and habitat loss have been the most serious pressures on gorilla numbers. CMS endorses UNEP's Great Apes Survival Partnership (GRASP), and promoted the negotiation and successful conclusion of the Agreement on the Conservation of Gorillas and Their Habitats, which seeks to emphasise transborder conservation programmes, and to develop gorilla ecotourism as a source of conservation and community income.

Addax, other antelope species and gazelles are key species in the biodiversity of the North African Sahelo-Saharan region. They have developed unique adaptation responses to the most arid environment. In addition to being a primary source of food they have historically played a major role in the livelihood of local communities. But due to severe man-made impacts on their habitats and excessive hunting they are rapidly declining. The magnificent Scimitar-horned oryx, still roaming the Sahel by the thousands in the 1970s, is now extinct in the wild. An Action Plan for six seriously endangered species, developed with active support of the Convention, recommends strong in situ conservation projects where the species still exist, particularly in Niger, while reintroducing and reinforcing some of the populations in the wild with captive-bred individuals, reducing mortality and enhancing international co-operation. This CMS program has benefit from a strong support from France and the European Union. With support of European and American zoos more than 50 antelopes were translocated to Tunisia in 2007. An agreement on the conservation of the Sahelo-Saharan megafauna is currently being negotiated, at the request of all 14 Range States.

African elephant populations have become extremely vulnerable, particularly in west Africa. The loss of 90% of their habitat and illegal killing are the primary causes. The Memorandum of Understanding therefore aims mainly to stop illegal killing and to reduce the rate of habitat loss of this, the world's largest land species.

Steppes and deserts are a favoured domain of activity for CMS. Arid zones, despite their relatively low species density, host a number of highly emblematic and remarkably adapted species. The conservation and restoration of the unique megafauna of the mountains and steppes in cold and temperate deserts and semi-deserts of Asia and Europe are essential for these exceptional habitats. The Central Eurasian arid-land concerted action covers Eurasian mammals, such as Bactrian camels, yaks, khulans, Snow leopards and gazelles. The objective of the program is to federate existing conservation projects to maximise international cooperation, to restore populations of all large mammals to viable levels, and to rehabilitate their habitats, throughout their range.

The Central Asian Bukhara deer, once numbered in great quantities, faces the threat of extinction as a result of human activities. Today, only a few hundred animals remain as a result of illegal hunting as well as artificial regulation of the water regimes in the river valleys where they live. A Memorandum of Understanding developed under CMS aims to save the species from the brink of extinction.

Until the early 1990s, more than one million Saiga antelopes used to roam the steppes and deserts of Eurasia. In recent decades, poaching for the Saiga's meat and horn, which is used in Chinese traditional medicine, has contributed to the decline of all Saiga populations. The aim of the Memorandum of Understanding is to reduce current exploitation levels and restore the population status of these nomads of the Central Asian steppes.

The CMS Agreement on the Conservation of Populations of European Bats (Eurobats) deals with 45 species known to occur in Europe. The most immediate threats to them nowadays derive from degradation of the places where they live, disturbance of roosting sites and certain insecticides and pesticides. European bats achieved new and improved legal protection standards for bats. The "European Bat Night" is a popular annual awareness-raising event celebrated all over Europe.

Monarch butterflies are found all around the world in temperate, sub-tropical to tropical areas. They are found in open habitats including meadows, fields, marshes and cleared roadsides. Not all populations are migratory, but those that are cover distances of up to 3,000km. The migration may take up to three generations to complete with the females laying their eggs along the way. The destruction of their habitat through the construction of roads, housing developments and agricultural expansion poses the greatest threat. Conservation measures thus focus on habitat restoration and protection.

Previous Pages:
African elephants (*Loxodonta africana*) gather around a watering pool, Namibia (Etosha National Park).

Opposite:
A herd of migrating wildebeest scramble to clear the water before the lurking crocodiles strike.

GORILLAS

CMS has signed the 2005 Kinshasa Declaration on the Conservation of Great Apes, described as the "Kyoto of the Great Apes" and is a partner of the Great Apes Survival Partnership (GRASP). An Agreement on the conservation of gorillas under CMS entered into force on 1 June 2008. The Agreement aims to bring together the governments of the ten Range States to guarantee the conservation of the populations of all gorilla species (Lowland as well as Mountain).

At the time of the Agreement's negotiation, the following subspecies were recognised: *Gorilla beringei beringei* (the mountain gorilla, pictured left in Rwanda); *Gorilla beringei graueri* (the eastern lowland gorilla); *Gorilla gorilla gorilla* (the western lowland gorilla); and *Gorilla gorilla diehli* (the Cross River gorilla).

The distribution range of these species includes these African countries: Angola, Cameroon, Central African Republic, Congo, Democratic Republic of Congo, Equatorial Guinea, Gabon, Nigeria, Rwanda and Uganda.

Commercial bushmeat hunting and for the exotic pet trade continue to present grave threats to the gorillas, as do habitat loss and fragmentation, civil unrest and war. Recent reports suggest that the Ebola virus is one likely cause of the recent disappearance of at least half the gorillas estimated to have lived in west-central Africa.

SAIGA ANTELOPE

Saiga tatarica

Saiga antelopes once numbered in the millions, creating an unforgettable spectacle as they raced across the steppe of Eurasia. In recent decades, poaching and habitat loss have led to a population drop of more than 90%, one of the most dramatic recent declines of any mammal.

The Range States are Kazakhstan, Russian Federation, Turkmenistan, Uzbekistan and Mongolia. The Saiga lives in the semi-desert steppe and arid grasslands of Central Asia. Because of their nomadic grazing, the Saiga herds play an important role in these ecosystems. Although some Saiga habitat has been degraded in the past, the current quality is generally good due to major reductions in livestock since the break-up of the Soviet Union.

There are four populations of *Saiga tatarica tatarica* in the Range States: the Ural, Ustiurt, Betpak-Dala and north-west Precaspian. A separate sub-species, *Saiga tatarica mongolica,* is found in Mongolia. Alarmingly, scientists estimate that less than 100,000 Saigas remain from a former population of nearly two million. Obstacles such as oil and gas pipelines can disrupt migration. Saiga antelopes are unique nomadic animals with bodies adapted to harsh and

unpredictable conditions. They are the size of goats with elephant-like snouts, which act as air filters, allowing them to breathe in cold, sandy environments. They are almost always travelling, moving across large distances for water or to escape predators such as wolves. Saigas can outrun natural predators — reaching speeds up to 80km an hour.

The IUCN lists this species as 'critically endangered', due to poaching (for meat and the male's horn for use in traditional medicine). International trade in Saiga horn has been regulated by CITES since 1995. Saiga antelopes are legally protected throughout their range and in China. Yet uncontrolled hunting remains the largest threat to the antelope. The Saiga's rapid decline started after the collapse of the Soviet Union, which had enforced strict control over Saiga hunting. Poverty drove local people to hunt Saigas unsustainably for their horns, meat and skins. The horns remain an extremely valuable, sought-after ingredient in traditional Asian medicine.

SAHELO-SAHARAN ANTELOPES

Six antelope species native to the Sahelo-Saharan region are listed in Appendix I of the Convention. These antelopes show a distinct physiological, morphological and behavioural ability to adapt to arid environments. Therefore, they are important species for maintaining biodiversity, particularly vegetative and predator communities, in the Sahelo-Saharan region. Due to over-hunting and degradation of habitat, Sahelo-Saharan antelopes have been in rapid decline for many decades. Hunted by man and other animals for food, they have historically played a major role in the culture and livelihood of local peoples of the region. But due to excessive hunting and poaching activities and severe habitat degradation over the past few decades Sahelo-Saharan antelopes have been in rapid decline and some of those species are extinct in the wild or in danger of extinction. The last confirmed sighting of a Scimitar-horned oryx in the wild was in Niger in 1988.

Species include Scimitar-horned oryx (*Oryx dammah*), Addax (*Addax nasomaculatus*), Dama gazelle (*Nanger dama*), Slender-horned gazelle (*Gazella leptoceros*), Cuvier's gazelle (*Gazella cuvieri*) and Dorcas gazelle (*Gazella dorcas*).

Above:
The Dama gazelle (*Nanger dama*) migrates across the Sahara in search of food in the dry season.

OPPOSITE:

SCIMITAR-HORNED ORYX
Oryx dammah

The Scimitar-horned oryx stands 1.4 metres tall at the shoulder and can weigh over 200kg. Both the male and female of this species have long, curved horns which can reach several feet in length. To retain fluid levels in the desert heat, the oryx only sweats when its body temperature reaches 46°C (114°F). The Scimitar-horned oryx's original habitats – it is now extinct in the wild – were chiefly arid plains and deserts, and secondly rocky hillsides and thick brush. It rarely entered true desert areas. Its range historically included the entire Saharan and Sahelian North Africa, between the Atlantic and the Nile. Captive-bred specimens have been reintroduced in Tunisia, Morocco and Senegal.

The oryx usually lived in herds of 20-40 led by a male. During migration, herds of up to 1,000 used to form. They migrated northwards during the wet season and south again in the dry season.

The IUCN Red Data List status for the Scimitar-horned oryx is 'extinct in the wild'. The principal cause was direct hunting, which increased with the advent of motorised vehicles and modern firearms, and was carried out by nomads, oil surveyors and military personnel both for sport and for meat and hides. Long-term climate change has rendered the oryx's habitat less suitable, and human activities have reduced tree cover and led to overgrazing by livestock of traditional oryx pasture.

LEFT:

BUKHARA DEER

Cervus elaphus bactrianus

The Bukhara deer's range states are Afghanistan, Kazakhstan, Tajikistan, Turkmenistan and Uzbekistan. Floodplain (riparian) forests, especially the Tigrovaja Balka in Tajikistan, are home to over 30% of the remaining population. Historically the species could be found over a wide area which included all the river valleys of Amudaria and Syrdaria, and their river catchment areas.

Now only a few hundred animals remain, scattered in a few small populations in limited areas. Estimates of the population are still under 1,000 but numbers have increased from the all-time low point of 350 in 1999.

The Bukhara deer, also known as the Bactrian deer, is a subspecies of red deer and is listed as 'vulnerable' on the IUCN Red Data List. Artificial regulation of the water regime, habitat destruction and illegal hunting are the main reasons for the Bukhara's alarming decline in numbers.

FOLLOWING PAGES:

WEST AFRICAN ELEPHANT

Loxodonta africana

The African elephant is the largest living land animal, the Savannah subspecies being larger than the Bush or Forest subspecies. The male weighs up to 7 tonnes and stands 4 metres tall; elephants weigh 120kg at birth after gestating for 22 months. The elephant was long hunted for the ivory of its tusks and elephant numbers dropped from millions to a low of 400,000 in 1989.

There is some debate among scientists about whether the two subspecies of African elephant currently recognised – the forest elephant (*Loxodonta africana cyclotis*) and the savannah elephant (*Loxodonta a. africana*) — are in fact two separate species. The forest elephant, with a smaller body, smaller ears and straighter tusks than the savannah subspecies, lives mainly in the rain forests of central and western Africa, while the savannah elephant's habitat is the grassy plains, woodlands, swamps and bushlands at altitudes ranging from sea level to high mountains.

In recent years, West African populations of the species have become extremely vulnerable. An estimated 90% of their range has been destroyed. This loss of habitat and illegal killing raise concerns about the future of the species. The elephants are also an important factor in savannah and forest ecosystems.

Elephant habitats include both humid forest and the arid Sahel. As humans move into these areas, elephants have less space and the number of human-elephant conflicts increases. Roads and railways also split the elephant range into isolated fragments. Two-thirds of these populations have fewer than 100 elephants, a problem since the larger the group, the better the chances of long-term survival.

There are approximately 500,000 African elephants left in the wild. Southern and eastern populations are generally thought to be stable. The size of the central population is difficult to estimate because of the nature of its habitat. The western population is fragmented in pockets containing relatively small groups.

The IUCN lists the African elephant as 'vulnerable'. Loss of habitat due to the expansion of human settlement and agricultural land is one pressure. Poaching remains a problem despite restrictions through CITES in the trade of ivory. Some elephants are illegally killed for bushmeat.

BATS

There are about 1,100 different species of bat in the world, representing nearly 25% of all mammal species. Within Europe, 45 species of bat have been identified. Some bats have been recorded as seasonally travelling from the Baltic states: Latvia, Lithuania, Estonia, and possibly even Finland, to Northern Spain or Italy.

Bats are the only mammals which can properly fly. To find their way around bats combine their acute sight with a complex, advanced system of "echolocation" which allows them to orient themselves by "bouncing" their sounds off objects in their vicinity, both to avoid collisions, and to find and hunt their prey.

In Europe, bats eat flies and moths and other insects. Some European bats also serve as pollinators and seed dispersers of many plants important to humans. There is only one fruit-eating bat in Europe (the Egyptian fruit bat that belongs to the sub-order *Megachiroptera*). Bats are not only the best insect-repellant, but they are also one of the best natural indicators of the health of the environment around us. This is because bats flourish best where the ecosystem is vibrant and stable.

All European bats are to a greater or lesser extent endangered with extinction. Some have even become extinct in certain countries. The reasons for this are mainly loss of roosts, loss of feeding areas and increased use of pesticides, both in agriculture and in the protection of building materials against pest action, which in turn poison the bats which consume them. A widespread misunderstanding and prejudice arising from ignorance about bats and their lives and habits needs also to be countered.

SCHREIBERS' BAT

Miniopterus schreibersii

Schreibers' bat (*Miniopterus schreibersii*) is found across a wide range extending from southern Europe to Japan and the Solomon Islands, the Philippines, northern Africa, and northern and eastern Australia. Caves, rocky clefts, culverts, caverns and galleries are often used as roosts. Studies of this species in India showed that the bats in any given area tend to concentrate in one large cave but that individuals also spend part of their time in secondary roosts within a radius of 45 miles (70km). Females reach reproductive maturity at the age of two — the gestation period is 8-9 months. The longest-lived Schreiber's bat ever recorded reached the age of 16 years old. It hunts after sunset and can reach speeds of 55kph. It seeks its prey (moths and beetles) in open country often far from the roost.

In Europe, they hibernate from November until the end of March. In Romania wintering colonies made up of as many as 10,000 individuals have been recorded. In the north of its range this species tends to migrate more, with the winter quarters lying approximately 100 kilometres further south. The longest recorded migration was over 800km.

Schreibers' bat is on the IUCN red list for low risk, 'near threatened' species. It is not on the CITES list, however this species has declined significantly in western Europe and possibly throughout the world. Colonies that had contained thousands of individuals have disappeared. Schreibers' bat is especially sensitive to disturbance and may disappear locally if frightened by human activity. Destruction

BROWN LONG-EARED BAT

Plecotus auritus

In the summer months Brown long-eared bats (*Plecotus auritus*) form a nursery from May onwards and these usually consist of between a dozen and thirty individuals. Groups of as many as 200 have been recorded. They prefer to roost in the roofs of buildings such as barns and churches, in crevices and timber, or on chimneys and gable ends. As adults, their woolly fur is light brown on their backs and a lighter, yellow colour on their stomachs – juveniles are grey. Their ears are three-quarters the length of the heads – and can reach 40 mm. Their long thumbs help distinguish them from similar species, such as the much rarer Grey long-eared bat. They usually weigh between 6 and 12 grammes.

They hibernate from November to the end of March, roosting in caves and similar habitats, either in crevices or on the open wall, often with their ears folded under their wings. They tend to roost singly or in very small groups. They are often found with other bat species.

The Brown Long-eared bat is found across much of Europe (including the British Isles and southern Scandinavia). It prefers open woodland, both deciduous and coniferous, valleys, orchards, parks and gardens.

The species is largely sedentary but populations cross international boundaries. The longest recorded movement for an individual was 90km. The species is widespread and abundant throughout Europe and is one of the three most common bat species in the UK, where its population was estimated at 200,000 in the 1990s.

Brown Long-eared bats have declined across Europe as a result of changing land use, leading to reduced quantity and quality of suitable foraging grounds. Increased use of pesticides has affected the species, as have the chemicals used to treat roof timbers.

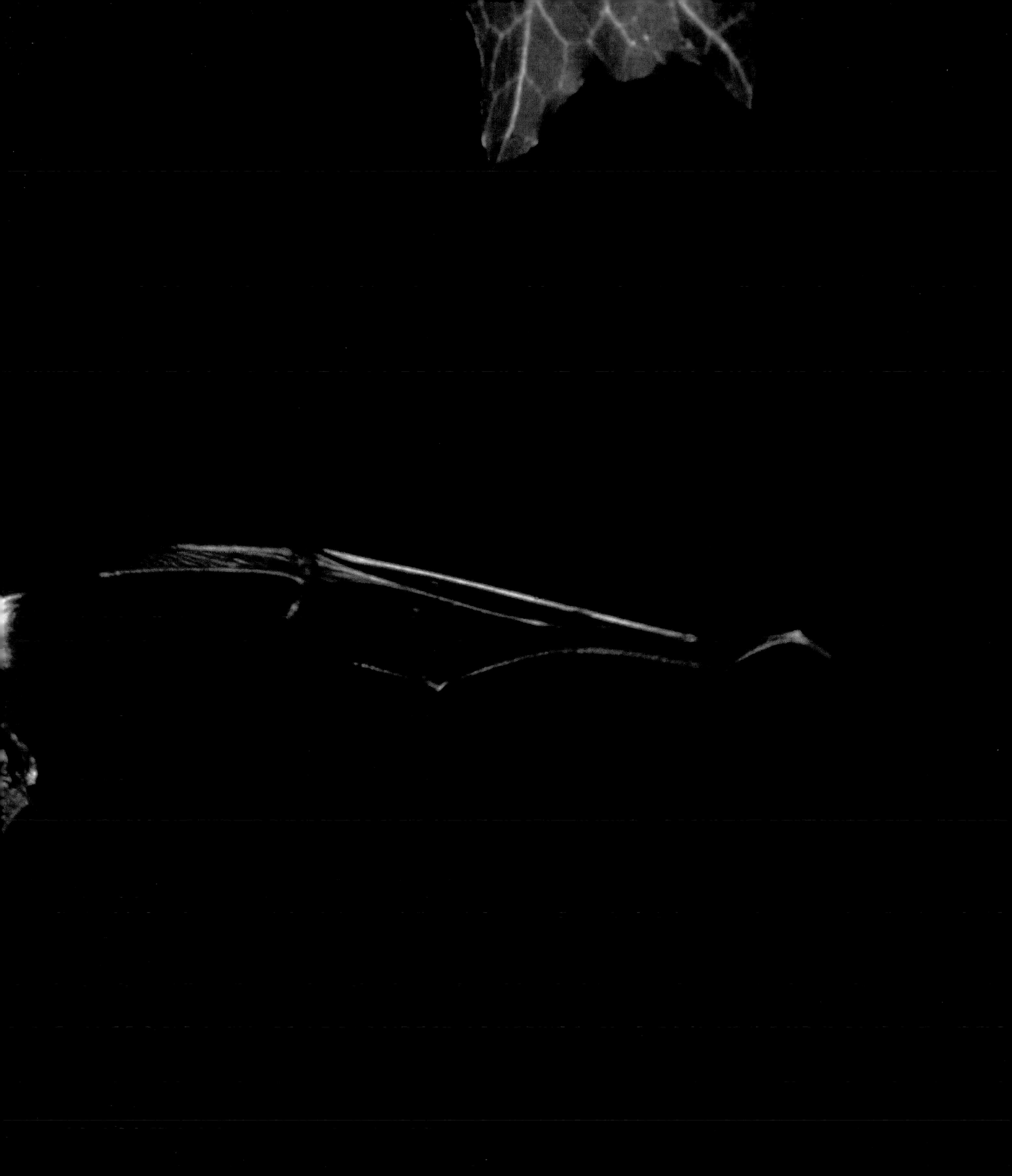

TIGER

The tiger is the largest of the four big cats of the genus *Panthera*. They can grow up to 4 metres in length and weigh 300kg – the larger subspecies such as the Siberian tiger (*Panthera tigris altaica*, pictured left) being of comparable size to extinct prehistoric felids.

Tigers are highly adaptable – with some subspecies living in the Siberian taiga and others in tropical swamps. Unlike the more sociable lion, tigers tend to be solitary and highly protective of their extensive territory. Females have a territory of 20 square kilometres while males' territories may be as large as 100 square kilometres. Most tigers live in areas with expanding human populations, which leads to conflicts and pressure on the animals' habitat.

Two of the eight subspecies of modern tiger are extinct and the remaining six are classified as 'endangered' or even 'critically endangered'. The main threats facing tigers are habitat destruction and fragmentation, direct persecution (hunting and poaching for the coat, for use in traditional medicine or to protect livestock) and because low numbers and low genetic diversity lead to inbreeding. All subspecies are legally protected.

The historical range of the tiger, which previously extended from the Tigris and the Euphrates (in modern day Iraq) and the Caucasus across large parts of Southern and Eastern Asia, has shrunk considerably. Tigers disappeared from Western Asia during the 19th century and their current range is highly fragmented from India in the West, China in the East, Indonesia in the South and the Amur River in Siberia in the North. Sumatra in Indonesia is the only island where tigers survive, the subspecies native to Bali and Java having become extinct.

The surviving subspecies are the (Royal) Bengal tiger (*Panthera tigris tigris*), found in the subtropical and tropical rainforests and mangroves of India and Bangladesh. Less than 1,500 of these tigers survive in the wild. The Corbett's or Indochinese tiger (*Panthera tigris corbetti*) is smaller and darker than the Bengal and occurs in Cambodia, China, Laos, Myanmar, Thailand and Vietnam. Only considered a separate subspecies since 2004, the Malayan tiger (*Panthera tigris jacksoni*), is only found in the south of the Malay Peninsula. The smallest subspecies is the Sumatran tiger (*Panthera tigris sumatrae*) found only on the island of Sumatra. Its size is an adaptation to the thick forest habitat. Only 400-500 remain in the wild, mainly resident in protected national parks. The largest subspecies is the Siberian or Amur tiger (*Panthera tigris altaica*) noted for its thick and paler coloured coat. Censuses in 1996 and 2005 put the numbers at between 450 and 500, while some recent research suggests that the now extinct Caspian tiger and the Siberian tiger were the same subspecies. The rarest tiger is the relatively small South China, Amoy or Xiamen tiger (*Panthera tigris amoyensis*). There have been no verified sightings in the wild since 1983.

In the wild, tigers' preferred prey consists mainly of large- and medium-sized animals. In India, tigers take Sambar, Gaur, Chital, Wild Boar, Nilgai and Water buffalo, but have been seen to eat smaller cats (such as leopards), snakes and even crocodiles. The Siberian tiger's diet is made up of wapiti, wild boar, deer and moose. Rhinoceroses and young elephants are known to have been taken.

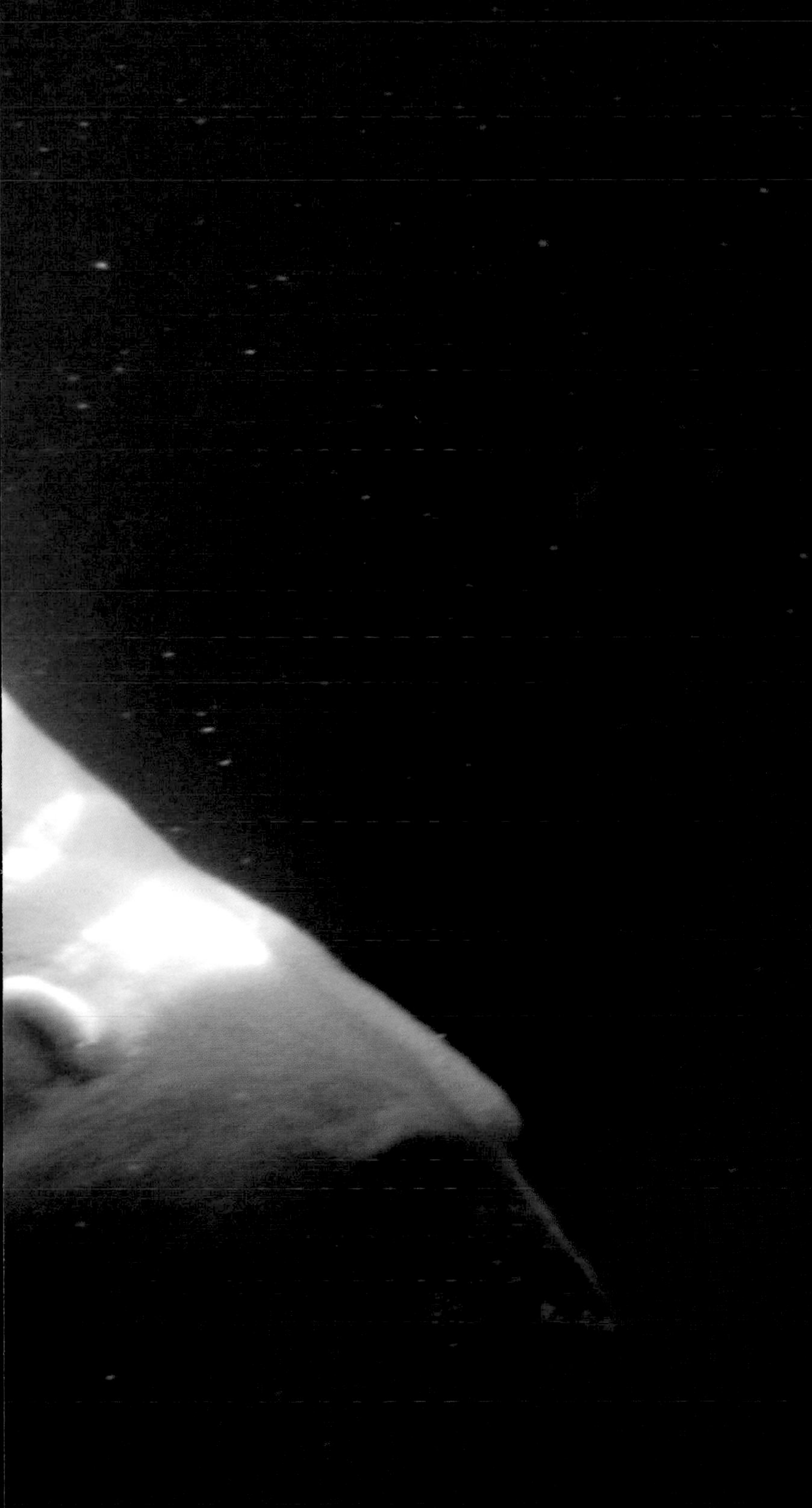

POLAR BEAR

Ursus maritimus

Polar bears are the largest terrestrial carnivores, with fully-grown males weighing as much as 680kg – twice as much as the females. In their Arctic habitat, they are apex predators (they are top of the food chain) and live on a diet mainly consisting of seals. They hunt by stealth, using their highly developed sense of smell, which can detect prey at distances of more than 1.5km. Polar bears sometimes tackle walruses, although the bear runs the risk of severe injury from the walrus's tusks, and Beluga whales. Both can weigh considerably more than the bear.

The Polar bear spends a large amount of time at sea, which accounts for its scientific name – *Ursus maritimus* – the maritime bear. They are strong swimmers and have been seen as far as 300km from the shore.

Polar bears are found in five countries: the USA (Alaska), Canada, the Russian Federation, Denmark (Greenland) and Norway (Svalbard), all of which are signatories to the International Agreement for the Conservation of Polar Bears concluded in 1973, committing these countries to cooperate on research and conservation programmes.

The Polar bear is listed as 'critically endangered' on the IUCN's Red List. Before 1976 it was classified in the 'least concern' category. The most pressing threat facing the species is global warming, which is causing the northern polar ice cap to shrink. This leads to starvation, as the bears are unable to hunt sufficiently and build up necessary reserves before the arctic summer. Other problems include pollutants, oil and gas exploitation and pressures from ill-managed wildlife tourism. Some NGOs consider the level of legal hunting to be unsustainable, while poaching adds further to the numbers taken.

Between 20,000 and 25,000 Polar bears remain in the wild. While a number of sub-populations have been identified, genetic research suggests that these intermingle. Polar bears lead an essentially solitary life but are not particularly territorial (certainly less so than other bear species) and usually avoid violent confrontations with each other.

When born, the cubs (usually two per litter) weigh less than 1kg. They remain in their underground dens for several weeks putting on weight very quickly, nourished by the mother's fat-rich milk. They emerge from the den having reached 10-15kg when they accompany their mother on the long journey to the seal hunting grounds where she can break a fast which might have lasted over six months.

RIGHT:

SNOW LEOPARD
Uncia uncia

The Snow leopard or Ounce (*Uncia uncia* or *Panthera uncia*) is a large cat whose range includes the rocky mountains of Central Asia and the rolling terrain of western China and Mongolia and the forests of Russia's Sayan mountains.

Snow leopards are smaller than the other big cats weighing between 27 and 54 kg (60 and 120 lb) and having a body length from 75 to 130 cm (30 to 50 in). Their tails are slightly shorter than their bodies.

Their build is ideal for the cold conditions in which they live. They have stocky bodies and thick fur, while their small, rounded ears help to minimise heat-loss. Their enlarged nasal cavity and strong chest help warm the cold air as they breathe in. The long tail is used for balance and, when curled around the animal's face at rest, serves as additional insulation. The short front legs and longer hind legs enhance the animal's agility in the steep and rugged terrain of its habitat.

The Snow leopard's range covers over one million square kilometres in the mountains of Central and South Asia. In 1972 the IUCN placed the Snow leopard on its Red List of Threatened Species in the 'globally endangered' category. It was added to Appendix I of CMS at the first Conference of the Parties in 1985. It is estimated that there are between 4,000 and 6,000 Snow leopards left in the wild with a further 600-700 in captivity.

The principal prey species are Blue sheep (*Pseudois nayaur*) and ibex (*Capra sibirica*) but it also takes marmot, pika, hares, other small rodents and game birds. Home ranges vary from 10 square kilometres to 40 square kilometres in Nepal where the density of prey species is relatively high to as much as 140 square kilometres in Mongolia's relatively open landscape.

The illegal trade in live specimens, parts and derivatives is one of the major threats to the survival of the species. Specimens are taken live to supply zoos and private collectors or killed as trophies for their pelts to make hats or coats, bones are used as substitutes for tiger derivatives in traditional medicine. The inaccessible and mountainous terrain makes law enforcement such as combating poachers very difficult. Collaboration between Range Countries offers a means for addressing this problem.

FOLLOWING PAGES:

WILD YAK
Bos mutus

The Wild yak is closely related to other species of the genus *Bos* including the banteng, gaur, the very rare kouprey and domestic cattle with which it can interbreed. The IUCN recognises both a wild and domestic form of yak, with taxonomic names of *Bos mutus* and *Bos grunniens* respectively, although some experts consider the former to be a subspecies of the latter.

Its Red Data List status is justified on the grounds of the 30% decline in the population over the past three decades, with a wild population estimated at 10,000 mature individuals compared with 15,000 in 1995.

The historic range of the yak includes Bhutan, Nepal, India and China, and possibly centuries ago in Kazakhstan, Mongolia and southern Russia, but no reliable records exist to support this. The species is now extinct in Bhutan and Nepal, while very few remain in India. In China there are three main populations – in the Chang Tang Reserve in Tibet, in Qinghai Province and in Xinjiang – and some further smaller isolated populations.

Poaching and encroachment by humans are the two most serious threats to the yak. While poaching, normally for the animal's meat, has declined, more intensive livestock farming has reduced the yak's habitat. Yaks have a low tolerance of disturbance and tend to move away from people and domestic livestock. Their main natural predator is the Tibetan wolf (*Canis lupus chanco*).

Interbreeding with domestic yaks is also a threat to the remaining wild populations, while there are suspicions that disease can be spread from livestock, which may have resulted in instances of poor reproductive success in recent years.

AFRICAN WILD DOG

Lycaon pictus

The African wild dog (*Lycaon pictus*) was formerly widely distributed across sub-Saharan Africa, occupying all habitats except the driest deserts and the wettest rain forest. Only 6% of the historic range still supports resident populations and Wild Dogs have all but disappeared from West Africa and have been greatly reduced in the north. The largest populations can be found in southern Africa, particularly northern Botswana, western Zimbabwe and eastern Namibia, as well as in Tanzania and Mozambique in East Africa.

Estimates made in 2003 and 2004 by IUCN suggested that slightly more than 7,500 wild dogs may remain in Africa, living in 742 packs and covering an area of over one million square kilometres.

The current distribution of the wild dog appears to be limited by factors such as human activities and the availability of prey. Loss of habitat is less of a problem, as they are a generalist species and are found in open plains, thicker bush and woodland, desert and montane habitats, only being absent from lowland forest. Habitat fragmentation, on the other hand, is a major problem because of the species' requirement for an extensive range.

Packs have home ranges varying in size from 150 square kilometres to 4,000 square kilometres, with the average lying somewhere between 600 and 800 square kilometres. Migration is not cyclical or seasonal as in the case of birds and antelope but seems rather to be determined by the need to avoid predators and find prey. In many cases, pack movements involve crossing national jurisdictional boundaries.

Although most wild dogs live in nominally protected areas, conflicts with humans are the principal direct threat to the species alongside infectious diseases. Habitat fragmentation has led to wild dogs predating on domestic livestock resulting in persecution by farmers. The proximity to domestic dogs has also facilitated the transmission of canine diseases. An outbreak of rabies eradicated one protected population and also thwarted efforts to reintroduce the species in two localities. Traffic accidents also take their toll as the wild dogs follow roads and have a tendency to rest on them.

The listing on Appendix II of CMS at the Conference of the Parties in December 2008 was the first formal international protection afforded to the wild dog. There is little consumptive use of wild dogs and they are not listed by CITES. The African wild dog has however been entered on the IUCN Red List as 'endangered'.

CHEETAH
Acinonyx jubatus

The cheetah (*Acinonyx jubatus*) is the fastest land animal, capable of speeds of 120 kph, which it can reach in three seconds. However, it can only maintain high speeds over 300-400 metres before becoming exhausted. Its small head and lightweight skull add to the animal's aerodynamics and its narrow body with its long, slender legs and specialised muscles are designed for speed. Semi-retractable claws add traction like the spikes on a sprinter's shoes. It is an atypical member of the cat family and therefore has been assigned its own genus, *Acinonyx*.

Five subspecies have been recognised although the differences between them are sometimes minimal. The five are *Acinonyx jubatus venaticus* found in North Africa and Asia; *A.j. hecki* in West Africa; *A.j. soemmeringii* in Central Africa; *A.j. raineyii* in East Africa and *A.j. jubatus* in Southern Africa.

The cheetah is primarily diurnal, possibly to reduce competition with other predators that operate at night. The litter sizes are larger than other big cats' possibly as a strategy to offset higher mortality rates caused by lions and hyenas.

Cheetahs are social animals and have been seen in groups as large as 19. Male and female siblings remain together for several months after becoming independent of the mother and male siblings often remain together in so-called "coalitions", which help to hold and defend territories.

Cheetahs take numerous species as prey including: Thompson's gazelle (*Gazella thompsoni*), impala (*Aepyceros melamus*), dik-dik (*Madoqua kirkii*), springbok (*Antidorcas marsupialis*), wildebeest (*Connochaetes taurinus*), hare (*Lepus capensis*) and Barbary sheep (*Ammotragus lervia*).

The cheetah was once one of the most widely distributed of the land mammals. By 1900, there were approximately 100,000 cheetahs left in 44 countries throughout Africa and Asia. In 1975, the total population was estimated at 30,000 with just 100 remaining in Iran. During the last century, cheetahs became extinct in large parts of their range. In Africa, free-ranging cheetahs are now found in 29 countries of North Africa, the Sahel, East and Southern Africa. In Asia, cheetahs are found only in Iran and possibly Pakistan. The effective breeding population is thought to be fewer than 10,000 with no sub-population counting more than 1,000 individuals.

Loss and fragmentation of habitat, depletion of prey species, low genetic diversity and infectious diseases are among the main threats to the cheetah's survival. Farmers also consider cheetahs to be a danger to their livestock and therefore kill them, although drought, other natural causes, theft and other predators are probably more to blame for farmers' losses. Cheetahs do not breed well in captivity and in nature reserves they suffer from competition from other predators such as lions and hyenas.

The IUCN Red List categorises the cheetah as 'vulnerable', while the Iranian population is considered critically endangered. The cheetah is also listed on Appendix I of CITES, although quotas have been set for export of limited numbers from Namibia, Zimbabwe and Botswana. The Ninth Conference of the Parties to CMS listed the cheetah on Appendix I, excluding the populations from the three countries with CITES export quotas.

MONARCH BUTTERFLY
Danaus plexippus

The Monarch butterfly (right and following pages) occurs in the USA, southern Canada, Central America, most of South America, some Mediterranean countries, the Canary Islands, Australia, Hawaii, Indonesia and many other Pacific Islands. Monarchs are found all around the world in sub-tropical to tropical areas. They are found in open habitats including meadows, fields, marshes and cleared roadsides. Their bright orange colouration probably signals to potential predators the fact that Monarch butterflies are poisonous.

Not all populations are migratory. In North America, each autumn, Monarch butterflies congregate in their thousands (even millions) in the south of Canada to migrate southwards through the USA to Mexico, covering distances of up to 3,000km. Some overwinter in Angangueo, Michoacan province, Mexico; others go to Cuba and California. The migration may take up to three generations to complete with the females laying their eggs along the way. Storms in Autumn 2006 blew many Monarch butterflies across the Atlantic — some were found in Britain.

The greatest threat to the Monarch butterfly is the destruction of its habitat through the building of new roads, housing developments and agricultural expansion. Also, the milkweed plant, on which the larvae feed exclusively, is considered a pest species because it is noxious and so it is often eradicated. The Monarch butterfly has no special status on the IUCN Red Data list, but IUCN has recognised the butterfly's winter migration as a "threatened phenomenon". Natural disasters such as severe storms in the Mexican wintering grounds have led to the destruction of the butterfly's habitat and to heavy reductions in numbers.

4

THREATS & CHALLENGES

By the very fact of their being migratory, these species risk falling between individual countries' efforts to protect them. As a result, and until relatively recently, migratory species have suffered from a multiplicity of mortal dangers more than sedentary creatures. Efforts by organisations such as the CMS are beginning to bear results. This chapter outlines those key threats that must in large measure be overcome if species are to be saved.

Invasive Alien Species

There are a number of natural ways in which a species can extend its range; seeds carried by currents or by birds for instance. The human contribution to the process, both by accident and design, has dramatically increased the number of instances and the consequences of some introductions have been devastating for native wildlife.

Rabbits introduced onto Laysan and Lisianski Islands in the Pacific during the early 1900s denuded the islands of vegetation and fierce sand storms buried nests and filled burrows. Within two decades populations of Black-footed Albatross (*Diomedea nigripes*) collapsed and three endemic land birds became extinct before the rabbits finally ate themselves to near extinction and the remaining few were killed.

Red Deer (*Cervus elaphus*) were introduced to Patagonia, and proved to be aggressively competitive. They were able to supplant the Chilean Huemul (*Hippocamelus bisulcus*) in areas where the two species lived side by side because of their greater tolerance of disturbances and adaptability regarding the use they made of the habitat.

Before they were eradicated between 1977 and 1980, feral cats (*Felis catus*) on Little Barrier Island (New Zealand) severely reduced numbers of the Black Petrel (*Procellaria parkinsoni*).

Alien species can also threaten native biodiversity through hybridisation. A lack of reproductive isolation between alien and native species can cause genetic swamping, loss of native genetic diversity and extirpation. The introduction of the North American Ruddy Duck (*Oxyura jamaicensis*) into Europe by collectors of exotic waterfowl led to one of the best-known cases of concern about an alien species in relation to conservation of a globally threatened native species, the White-Headed Duck (*Oxyura leucocephala*).

Plants too can cause ecological problems. Rampant plants like the Japanese Knotweed (*Fallopia japonica*) can take over swathes of land, literally strangling out all competition and depriving animals of habitat and food. Like the Ruddy Duck, Japanese Knotweed was deliberately introduced to Europe because of its ornamental qualities.

CMS has undertaken a study based on but not limited to an analysis of threats posed by invasive alien species to migratory animals listed on the Convention's annexes and the effects, real and potential that they have. The study also examined prevention and control mechanisms including measures already being implemented and others under consideration.

Membership of the Convention commits Parties to taking measures to prevent exotic species, as alien and invasive species can also be described, from endangering the migratory species listed on Appendix I.

While the best guard against alien species is to maintain vigilance to prevent their introduction, this is often easier said than done. Effective control and eradication programmes have proved possible in a number of instances.

Key facts:

The main threats posed by invasive alien species:

- Competition with native species
- Detrimental impacts on habitat
- Direct predation on adults, young and/or eggs
- Hybridisation with the native species
- Diseases by pathogens and parasites

Bycatch

Bycatch, the accidental capture of a non-target species in fisheries, is both a common and universal phenomenon. Between a quarter and a fifth of all fish caught across the world is simply thrown overboard – that is the equivalent of 20 million tonnes of marine life discarded every year. Trawls, seines, hooks and lines, gillnets and driftnets and even lines of pots and creels take their toll on all sorts of animals – marine mammals, sea birds, turtles and sharks. Worst affected are long-lived, slow breeding species like cetaceans, seals, turtles and albatrosses. Indeed 19 of 21 species of albatross are threatened with extinction, and the primary threat they face comes from long-line fisheries.

Moreover, it is not just the species that suffer; entire marine ecosystems are damaged as they lose an important element of their structure. In the face of this serious threat, CMS has taken a lead and its Parties have endorsed resolutions and recommendations at the last four Conferences (Cape Town 1999, Bonn 2002, Nairobi 2005 and Rome 2008) calling for immediate action by the international community to address the problem and improve fishing practices to reduce the unnecessary death of so many non-target species. In addition, there are several CMS-related Agreements and Memoranda of Understanding dedicated to species for which bycatch is a major issue.

It is only recently that the extent of the problem of bycatch has become apparent. Our knowledge is improving as more data are collected and analysed and coverage by observers of fishing fleets increases. But the data obtained paints a gloomy picture as the conservation status of key species such as the Macquarie Island Wandering Albatross (*Diomedea exulans*) and Amsterdam Albatross (*Diomedea amsterdamensis*) remains alarming—both are near extinction. Another cause for concern is the fate of marine turtles. Across their entire migratory range bycatch is a problem but at least global action is now being taken. It is estimated that an annual average of 6,000 Harbour Porpoises (*Phocoena phocoena*) have died as a result of bycatch over the past decade in North Sea fisheries alone. Losses of this magnitude are unsustainable and populations will only recover when bycatch levels fall drastically.

Other species face similar threats, but have yet to attract the same level of attention to their plight. The total annual of marine mammal bycatch is thought to be in excess of 300,000 individuals. For some species, like the Vaquita (*Phocoena sinus*) which is only found in the Gulf of California, extinction looms; for others, like the Irrawaddy Dolphin (*Orcaella brevirostris*), the extent of bycatch has yet to be ascertained and no remedial actions are being taken.

The Food and Agriculture Organisation (FAO) has been instrumental in negotiating International Plans of Action aiming to reduce bycatch levels of sharks and sea birds. Innovative fishing techniques are the subject of trials in the Southern Hemisphere to

Right and Overleaf:
The albatross has suffered greatly from inappropriate fishing practices. Nineteen of the 21 species are threatened with extinction. Albatrosses have the largest wingspan of all birds. The Black-footed albatross (*Diomedea nigripes*) is pictured on the right and overleaf, a pair of Black-browed albatross (*Diomedea melanophris*), alighted at their nesting site on the Falkland Islands.

reduce albatross and petrel losses. In several fisheries in the Indian Ocean and South Pacific it is now required to fit turtle excluder devices. A start is being made to addressing bycatch with stricter regulations being enforced in many regions, but still more needs to be done. The CMS Recommendation 7.2 adopted in 2002 calls on Parties to compile information to assess the impact of bycatch on migratory species and take action regarding fishing activities within their control.

Key facts:

Effective bycatch reduction requires:

- Appropriate management frameworks to ensure that conservation objectives are identified and appropriate action taken to meet them.
- Responses which are adapted to the species and fisheries involved.
- Solutions which can respond to a growing world fishing fleet equipped with more efficient gear.
- A dedicated expert on bycatch was appointed to the CMS Scientific Council in 2006.

Electrocution

Recent figures compiled by experts for NABU (Naturschutzbund, a leading German conservation NGO) show how great the risk of bird electrocution is in Central and Eastern European countries. Looking at Estonia, Poland, the Czech Republic, Hungary, Slovenia and Croatia, as many as 42 bird species listed in Appendices I and II of CMS are threatened due to power poles that have yet to be fitted with the latest safety devices. Twenty-two of the affected species are already classified as 'critically endangered'.

When a bird's wings bridge the gap between wires carrying different voltages, a short circuit occurs. Electricity passes through the bird's body, causing severe burns and in the worst cases fatal paralysis. More common are ground faults, when the bird bridges the gap between the wire carrying the electric current and the pole supporting the wire. In humid weather conditions the risk of electric sparks or electric arcs increases. Birds can also injure or kill themselves when they collide with power masts or overhead cables. Further examples from Kazakhstan in Central Asia show the horrendous effects that poorly designed power poles can have. In a nature reserve on Lake Tengiz, numerous birds, including 200 kestrels, 48 Steppe Eagles (*Aquila nipalensis*), two Eastern Imperial Eagles (*Aquila imperialis*), one White-tailed Eagle (*Haliaeetus albicilla*) and one Black Vulture (*Aegypius monachus*) were recorded killed by electrocution along an eleven-kilometre medium voltage overhead powerline in the month of October 2000 alone.

Studies show that in many regions, electrocution poses one of the greatest risks to large birds and their populations. This is the case for the Eagle Owl (*Bubo bubo*) in Norway and for Bonelli's Eagle (*Hieraaetus fasciatus*) in Spain.

As long ago as 1913 at the Third German Bird Conservation Conference, an engineer, Hermann Hähnle, gave a talk entitled 'Electricity and Bird Protection', pointing out the disastrous consequences even then of electrocution of birds. He concluded that, "electricity utility companies should be required to provide comprehensive protection for wild birds so that, if accidents occur, mitigation measures can be put to work at once".

There is no good reason why any birds should be killed or injured due to electrical structures. Medium voltage power lines can be laid underground, and some German power companies such as E.ON Energie and EWE AG Lower Saxony have implemented this policy for several years.

Regulations and guidelines have been drawn up to ensure safer design of structures. Germany's Federal Nature Conservation Act of April 2002 requires newly erected power poles and technical hardware to be constructed to exclude the possibility of bird electrocution. Mitigating measures are to be undertaken on existing power poles and technical hardware in the medium voltage range within the next ten years. CMS is undertaking a project with NABU aimed at disseminating guidelines on the construction of safe poles.

Key facts:

CMS Parties at COP7 in 2002 passed a Resolution calling for:

- Safe construction through planning regulations.
- The neutralisation of existing pylons.
- Encouragement of power operators to adopt bird-friendly designs and practice.

NABU has produced a booklet: *Caution: Electrocution! Suggested Practices for Bird Protection on Power Lines*. The booklet is recommended by CMS, BirdLife International and EURONATUR.

Habitat Loss

One of the main threats facing endangered migratory animals is habitat loss. This problem manifests itself in a number of ways:

- direct loss of habitat and species – from agricultural improvements, urban development, mineral extraction and afforestation.
- fragmentation – the splitting up of continuous blocks of habitat and species into disconnected pockets.
- degradation – occurs when a habitat is no longer managed in an appropriate manner, for example: overgrazing on upland moors, undergrazing on lowland heaths and draining of wetlands.

Global Warming – the Arctic hunting grounds of the polar bear are melting. Warmer waters away from the Tropics mean that cold-water species face more competition for food as the range of other species spreads and their own preferred habitat becomes more scarce. The loss of the tundra habitat in Siberia as a result of the northward shift of forests is having a severe impact on many species which rely on the Arctic tundra for breeding. One example of such a species is the Spoon-billed sandpiper (*Eurynorhynchus pygmeus*), which is listed on Appendix I of CMS and has been identified for concerted action.

Desertification – the expansion of deserts such as the Sahara is making them far more formidable barriers for migratory species; some birds have difficulty building up sufficient food reserves to make a successful crossing.

Deforestation – we often hear that an area of pristine forest the size of hundreds of sports fields or a small country has been destroyed by fire (accidental or intentional) or felling. Many rare species have been lost before scientists have had the chance to record them.

Urbanisation – the world's urban population exceeded 50% for the first time in 2007 and is projected to reach 5 billion or 61% by 2030. Cities are encroaching on the countryside. While some species can adapt and thrive, others are displaced and their numbers decline.

Agriculture – habitats are lost as land is converted to agricultural use, often with increased use of chemical fertilisers and pesticides. With less natural habitat to meet their needs, animals often seek refuge close to human settlements, often leading to conflicts with people.

Irrigation – flood defences aimed at protecting human settlements and agricultural land damage some natural habitats which require periodic flooding to retain the features required by some species. The habitat favoured by the Bukhara Deer (*Cervus elaphus bactrianus*) was affected by artificial irrigation.

Pollution – use of pesticides and chemical fertilisers has had an adverse effect on many species. Rain washes these pollutants into aquatic habitats poisoning the animals or the prey or plants upon which they feed. Industrial waste, discharges from ships and oil spills add to the problem. Autopsies of stranded cetaceans often reveal high levels of heavy metal contamination.

Key facts:

- An area of unspoilt land larger than North America is likely to be damaged by human activity in the next 30 years.
- Logging and mining each affect one in six endangered species, grazing one in five, water development nearly a third, recreation affects about a quarter.
- For migrant bird populations, a decline of close to 40% is directly linked to habitat destruction.

Wind Turbines

Electricity generation by wind turbines is controversial, and objective assessments of this technology's impact on migratory species are rare. Recently, the Bern Convention (the Council of Europe) commissioned a report from BirdLife International on how wind turbines affect birds and how to minimise the risks. The principal dangers for birds are collision, disturbance leading to birds leaving otherwise ideal habitat and some loss of land needed to build and service the wind farms. Noise pollution is the main problem arising from coastal turbines for cetaceans.

Right:
Polar bears (*Ursus maritimus*) migrate south as the Arctic winter tightens its grip. In some areas, melting ice has dramatically affected the bears' traditional routes, leaving the animals remote from their prey.

There have been well publicised examples of collisions with wind turbines leading to horrific numbers of deaths of birds. Even small increases in mortality rates due to wind farms can have significant impacts on the population of large, long-lived species which reach breeding age later and rear small clutches. Collisions with turbines are more likely to occur during abnormal wind conditions, rain and fog. The often remote and inaccessible location of many wind farms makes it difficult to record true mortality rates, but thermal imagery devices can help.

The extent of disturbance caused to birds depends on a number of factors: the size of the wind farm, the amount of habitat lost to construction and service infrastructure, the noise of the rotating blades and increased human presence during maintenance. The impact is less where adequate alternative habitat is within easy reach. The size and design of the installation (i.e. the spacing between each turbine) is important. Where ecological corridors between feeding, breeding and roosting sites are left intact, bird populations are generally not adversely affected.

On the basis of the precautionary principle, conservationists advocate that wind farm development should not be permitted in or near nationally or internationally designated sites (e.g. Special Protection Areas in the E.U.'s Natura 2000 network). Close monitoring before, during and after construction of pilot projects and dialogue between conservation interests and developers should contribute to solutions that provide clean energy without demanding too high a cost from wildlife and precious natural habitats.

In conjunction with the European Cetacean Society, CMS convened a workshop to discuss the impact of wind farms on cetaceans. The disturbance caused by turbines in operation was thought to be less of a problem than noise during construction. Another factor was the increased shipping traffic necessitated by maintenance work. The workshop considered best practice guidance for the construction period, including conducting visual and acoustic surveys to ensure that no cetaceans were present during noisy activities. Studies are needed to establish baseline data and to monitor the longer term effects of wind farms on cetacean populations.

Key facts:

- A German government sponsored resolution at the CMS COP7 in Bonn in 2005 called for an evaluation of the threat posed by turbines to migratory species, especially birds, and for environmental impact assessments to be carried out before permitting construction of wind farms in sensitive areas.
- The local authority in Altamont, California, negotiated a deal with energy providers and conservationists under which wind farm operators agreed to introduce measures to cut raptor collisions including the removal of some of the deadliest turbines and painting some blades as a deterrent.

Right:
The recent appearance of wind turbines – often in large numbers – both on- and off-shore poses a potentially major threat to migrating birds.

Oilspills

Of all man-made disasters, oil spills are among the most common and can, over localised areas, be one of the most destructive to wildlife, killing the animals and poisoning their habitats. Here is a list of some of the most recent major incidents:

March 2009: *Pacific Adventurer* off the coast of Queensland, Australia, affecting 60km of beaches.

December 2007: *Hebei Spirit* spills 10,000 tons off South Korea's west coast.

November 2007: a storm in the Strait of Kerch between the Sea of Azov and the Black Sea damaged two tankers and sank four other vessels.

July 2006: the oil storage unit at Jiyyeh, 30 km south of Beirut damaged in hostilities causing spillage affecting the coasts of Lebanon and Syria.

November 2002: the vessel *Prestige* carrying 20 million gallons (70,000 metric tons) of fuel oil broke up off the Spanish coast.

January 2001: the vessel *Jessica* spilled 175,000 gallons of diesel and bunker oil into the sea off the Galapagos Islands.

June 2000: 1,400 tonnes of heavy fuel oil leaked from the bulk carrier *Treasure* off Cape Town, affecting penguins on Dassen and Robben Islands.

January 2000: A ruptured pipeline spewed about 340,000 gallons of heavy oil into Guanabara Bay, Rio de Janeiro.

February 1996: The *Sea Empress* hit rocks near Milford Haven, Wales, spilling 72,000 tonnes of oil.

January 1993: The *Braer* sank off the Shetland Islands spilling 85,000 tonnes of light crude oil.

Despite greater vigilance, faster and better equipped response teams, tougher regulations and improved ship design, the threat of oil pollution from accidents or deliberate discharge cannot be ignored. The 20,000 birds found dead or dying on the Spanish coast after the *Prestige* disaster 2002 were just the start. The final toll from this one incident was more likely to have been counted in the 100,000s.

Shipping accidents leading to high losses among wildlife are still a regular occurrence across the world and Europe's busy sea-lanes are no exception. Deliberate discharges of noxious waste from ships account for the deaths of yet more birds, sea mammals, fish and reptiles.

Waders and seabirds like auks, sea ducks and other diving birds which feed in the sea are worst affected. When oil sticks to birds' plumage, the feathers lose their insulating properties and the birds die of cold. Marine mammals are susceptible to oil contamination too. One problem is that the animals try to clean themselves and ingest the oil which poisons them. Unable to stay in the water, they move onto land where they succumb to poisoning or hypothermia.

Although rescue efforts save some oiled birds, most unfortunately die. Insufficient baseline data makes it difficult to assess long term effects of oil spills, but accidents occurring at sensitive times of the year could claim large numbers of adult birds and have devastating effects on local populations. An accident in the mouth of northern Germany's River Elbe, a busy shipping lane, in the late summer is a particularly alarming prospect as the mudflats in the area support virtually the entire north west European Shelduck (*Tadorna tadorna*) population. On the other hand, research into breeding responses after the *Braer* oil spill in 1993 tended to indicate that some bird populations made a swift recovery.

Left:
Oil slicks continue to claim the lives of many seabirds which, unable to swim, starve to death. The Jackass penguin (*Spheniscus demersus*) pictured opposite fell victim to an oil spillage off the coast of South Africa.

Key facts:

- After the oil storage depot at Jiyyeh was damaged in the conflict of summer 2006, CMS offered its expert advice to the Lebanese and Syrian authorities. This was part of a wider UN effort to mitigate the environmental effects of the spillage.

Climate Change

Our climate is changing – the Earth's temperature and sea levels are rising, rain patterns are altering, and extreme weather is occurring more often. Conservationists are facing new challenges. In conjunction with the UK's DEFRA, CMS has produced a publication entitled *Migratory Species and Climate Change: Impacts of a Changing Environment on Wild Animals*, explaining the problems before us.

Migration – Abnormal storms have blown Monarch Butterflies (*Danaus plexippus*) across the Atlantic from America to the UK. Weaker tailwinds in Siberia result in fewer Bewick's Swans (*Cygnus columbianus*) reaching traditional wintering grounds in Europe. Winters are often so mild that cranes delay their southbound flight by weeks and sometimes do not migrate at all.

Habitat – The range of species is moving towards the poles and higher elevations. Cold-water species are facing increased competition as the seas nearer the poles warm up. Exotic fish like the anchovy are now found in the North Sea, while bird species once confined to the arid Sahara are finding suitable conditions on the northern Mediterranean.

Feeding – Polar Bears (*Ursus maritimus*) may not adapt fast enough to changing conditions affecting the habitat of their prey species. The growth of deserts is making it more difficult for migratory species to cross these barriers, as the animals must eat more food to survive the journey but have fewer opportunities to do so.

Breeding & Nesting – All the Galapagos Fur Seal pups (*Arctocephalus galapagoensis*) born in 1982 were lost as a result of the El Niño Southern Oscillation. The same phenomenon has affected Green Turtles (*Chelonia mydas*) migrating to their breeding grounds. Higher sea levels will also erode breeding beaches. Bats are waking from hibernation early affecting the females' reproductive cycle.

Resting – Loss of sea ice is affecting the Ringed Seals (*Pusa hispida*), Bearded Seals (*Erignathus barbatus*) and Walruses (*Odobenus rosmarus*) that use ice floes to rest, moult and give birth. The Lesser White-

fronted Goose (*Anser erythropus*) is particularly vulnerable as it relies on a small number of discrete stopover sites.

Disease – Tumours in Green Turtles (*Chelonia mydas*) grow faster in warmer waters. Other diseases and parasites thrive in higher temperatures. Algal blooms contribute to epizootic episodes, while viral outbreaks have reduced the effectiveness of animals' immune systems, leading to mass die-offs.

Feminisation – Hotter nesting beaches are affecting both the survival of turtle eggs and the gender ratios of hatchlings. Eggs need temperatures of 25-32°C to incubate successfully. At the lower end of the range, predominantly male hatchlings are produced; at the higher end mainly female. With ratios of 1 male to 4 females, adverse effects on populations arise.

Key facts:

In 1997, the fifth CMS Conference of Parties COP5 passed a Recommendation on climate change, followed by Resolutions in 2005 and 2008 calling for, inter alia:

- More scientific research into the effects on migratory species and to revise Range State lists as species' movements change.
- Greater collaboration and stronger links with other environmental conventions dealing with climate change.
- The development of adaptation measures to counteract the effects.

Wildlife Watching

CMS has produced an illustrated publication in conjunction with the German travel firm TUI: *Wildlife Watching and Tourism: a Study on the Benefits and Risks of a Fast growing Tourism Activity and its Impacts on Species*.

However, to achieve these benefits, wildlife watching tourism needs to be carefully planned and managed by government agencies, the tourism sector and conservation practitioners. With rapidly growing demand from tourists for wildlife watching activities, controls are also needed to prevent adverse effects on wildlife and local communities.

Economic and Social Benefits – It is estimated that in the USA the direct expenditure on wildlife-watching was US$32 billion in 2001, including US$7.5 billion on food, transport and accommodation. One in five US citizens lists birdwatching as one of their pastimes. Between 20% and 40% of international tourists have some interest in wildlife-watching, from casual observation through going on a specific wildlife-related excursion to an entire trip dedicated to wildlife watching.

Conservation Benefits – As the actual revenues from wildlife watching are large, there is considerable potential to channel some of this money

Left:
Deforestation contributes approximately 20 per cent of annual greenhouse gas emissions. The loss of the world's forests is already thought to have had a significant impact on the planet's weather patterns, with direct consequences for many migratory species.

towards the conservation of the species being observed. For example, the 'Projeto Tamar' in Brazil has promoted the conservation of turtles along the coast, and by protecting hatcheries, the number of young turtles reaching the sea reached 600,000 in 2003 alone. Mountain Gorilla numbers are rising fastest in those populations in the Democratic Republic of Congo, Rwanda and Uganda where tourists visit most regularly. The picture is similar with the whale populations off the Valdes Peninsula in Argentina.

Examples of Wildlife Watching – Monarch Butterflies (USA, Canada and Mexico); snorkelling with sharks (Indonesia, Seychelles, Red Sea and Caribbean); observing turtles in Brazil, Mexico, Cape Verde, South Africa, Sri Lanka and Indonesia; albatrosses – visiting breeding colonies in New Zealand; cranes – observing birds in Germany and the USA; penguins (Antarctica, Argentina, South Africa, Australia); gorillas (national parks on the borders of the Democratic Republic of the Congo, Uganda and Rwanda).

Risks – stress to animals (some, like cranes, do not adapt well to human presence); disease – great apes can catch human illnesses; habitat degradation and disturbance to natural habitats by human activity and building tourist infrastructure; economic overdependence on tourism and 'Boom and Bust' leading to superfluous facilities.

Key facts:

- In the year 2000 alone, 1.5 million visitors to Kenya, Uganda and the United Republic of Tanzania spent over US$1,000 million.
- The number of international tourists is projected to reach 1,600,000,000 in the year 2020.
- Key questions in Wildlife Watching Tourism are: Can tourism be managed in a way which is compatible with the needs of the species and their habitats? Is there a market for such tourism? How would local communities benefit?

Right:
A lone African bull elephant (*Loxodonta africana*) patrols his territory.

APPENDICES

Left:
Magpie geese (*Anseranas semipalmata*) on migration in their characteristic flight formation at Kakadu National Park, Australia.

Pages 158-159:
Canada geese (*Branta canadensis*) on migration at twilight over Maryland, Florida.

Appendix I: List of Common Names CMS Appendices I & II – March 2009

App	Taxon	English	Français	Español	Deutsch	Português
I	*Acinonyx jubatus*	Cheetah	Guépard	Guepardo	Gepard	Guepardo
II	*Acipenser baerii baicalensis*	Baikal Sturgeon	Esturgeon sibérien (Baïkal)	Esturión del baikal	Baikal-Stör	Esturjão do Baikal
II	*Acipenser fulvescens*	Lake Sturgeon	Esturgeon lacustre	Esturión Lacustre	Roter Stör	Esturjão do Lago
II	*Acipenser gueldenstaedtii*	Russian Sturgeon, Ossetra	Esturgeon de Danube, oscietre	Esturión Ruso	Waxdick	Esturjão-russo
II	*Acipenser medirostris*	Green Sturgeon	Esturgeon vert	Esturión verde	Grüner Stör	Esturjão-verde
II	*Acipenser mikadoi*	Sakhalin Sturgeon	Esturgeon des Sakalines	Esturión de Sakhalin	Sachalinstör	
II	*Acipenser naccarii*	Adriatic Sturgeon, Italian Sturgeon	Esturgeon de l'Adriatique	Esturión del Adriático	Adria-Stör	Esturjão do Adriático
II	*Acipenser nudiventris*	Ship Sturgeon, Spiny Sturgeon	Esturgeon à barbillons frangés	Esturión Barba de Flecos, Esturión de Flancos	Glattdick	Esturjão-ventre-nu
II	*Acipenser persicus*	Persian Sturgeon	Osciètre	Esturión persa	Persischer Stör	Esturjão Persa
II	*Acipenser ruthenus*	Sterlet	Sterlet ou estrugeon de Sibérie	Sterlet	Sterlet, Sterlett	Esterlete
II	*Acipenser schrenckii*	Amur Sturgeon		Esturión de Amur	Amurstör	Esturjão do Rio Amur
II	*Acipenser sinensis*	Chinese Sturgeon		Esturión Chino	Chinesischer Stör	Esturjão Chinês
II	*Acipenser stellatus*	Stellate Sturgeon, Sevruga, Star Sturgeon	Esturgeon étoilé	Esturión Estrellado, Sevruga	Sternhausen	Esturjão Estrelado
I/II	*Acipenser sturio*	Common Sturgeon, Atlantic Sturgeon, Baltic Sturgeon, German Sturgeon	Esturgeon commun, Esturgeon d'Europe occidentale	Esturión Común, Esturión Atlántico	Gemeiner Stör, Baltischer Stör	Esturjão Comun
I/II	*Acrocephalus griseldis*	Basra Reed-warbler	Rousserolle d'Irak	Carricero de Basra	Basrarohrsänger	Felosa do Iraque
I/II	*Acrocephalus paludicola*	Aquatic Warbler	Phragmite aquatique	Carricerín	Seggenrohrsänger	Felosa aquática
I	*Acrocephalus sorghophilus*	Streaked Reed-warbler	Rousserolle sorghophile	Carricerín de Anteojos	Seggensänger, Hirserohrsänger	
I	*Addax nasomaculatus*	Addax	Addax au nez tacheté	Adax	Mendesantilope	Adax
II	*Aenigmatolimnas marginalis*	Striped Crake	Marouette rayée	Polluela Culirroja	Graukehl-Sumpfhuhn	Franga-d'agua-estriada
I/II	*Agelaius flavus*	Saffron-cowled Blackbird	Carouge safran	Tordo Amarillo	Dragón Gilbstarling	Veste-Amarela
I/II	*Alectrurus tricolor*	Cock-tailed Tyrant	Moucherolle à queue large	Yetapá chico	Rotkehl-Schleppentyrann	Galito
I/II	*Alectrurus risora*	Strange-tailed Tyrant	Moucherolle petit-coq	Yetapá de Collar, Tijereta de las Pajas	Hahnenschwanz-tyrann	Tesoura-do-campo
II	*Alopochen aegyptiacus*	Egyptian Goose	Oie d'Egypte, Ouette d'Egypte	Ganso del Nilo	Nilgans	Ganso do Egito
II	*Amazona tucumana*	Tucuman Amazon	Amazone de Tucuman	Loro Alisero	Tucuman-Amazone	Papagaio Tucumã
II	*Ammotragus lervia*	Barbary sheep	Mouflon à manchettes	Muflón de Berbería	Mähnenspringer, Mähnenschaf	Carneiro-da-Barbária, Arruí
II	*Anas acuta*	Northern Pintail	Canard pilet	Anade Rabudo	Spießente	Marreca-arrebio
II	*Anas capensis*	Cape Teal	Sarcelle du Cap, Canard du Cap	Cerceta del Cabo	Fahlente, Kapente	Marreco-do-cabo
II	*Anas clypeata*	Northern Shoveler	Canard souchet	Cuchara Común	Löffelente	Pato-trombeteiro
II	*Anas crecca*	Common Teal	Sarcelle d'hiver	Cerceta Común	Krickente	Marrequinha-comun

App	Taxon	English	Français	Español	Deutsch	Português
II	*Anas erythrorhyncha*	Red-billed Duck	Canard à bec rouge	Anade Piquirrojo	Rotschnabelente	Marreco-de-bicovermelho
I	*Anas formosa*	Baikal Teal	Sarcelle élégante, Canard de Formose	Cerceta del Baikal	Baikalente	Pato de Baikal
II	*Anas hottentota*	Hottentot Teal	Sarcelle hottentote	Cerceta Hotentote	Hottentottenente	Marreco-hotentote
II	*Anas penelope*	Eurasian Wigeon	Canard siffleur	Silbón Europeo	Pfeifente	Piadeira
II	*Anas platyrhynchos*	Mallard	Canard colvert	Anade Azulón	Stockente	Pato-real
II	*Anas querquedula*	Garganey	Sarcelle d'été	Cerceta Carretona	Knäckente	Marreco
II	*Anas strepera*	Gadwall	Canard chipeau	Anade Friso	Schnatterente	Frisada
II	*Anas undulata*	Yellow-billed Duck	Canard à bec jaune	Anade Picolimón	Gelbschnabelente	Pato-de-bico-amarelo
II	*Anser albifrons*	Greater White-fronted Goose	Oie rieuse	Ansar Careto Grande	Blässgans	Ganso-grande-de-testabranca
II	*Anser anser*	Greylag Goose	Oie cendrée	Ansar Común	Graugans	Ganso-comum, Gansobravo
II	*Anser brachyrhynchus*	Pink-footed Goose	Oie à bec court	Ansar Piquicorto	Kurzschnabelgans	Ganso-de-bico-curto
I/II	*Anser cygnoides*	Swan Goose	Oie cygnoide	Ansar Cisnal	Schwanengans	Ganso-africano
I/II	*Anser erythropus*	Lesser White-fronted Goose	Oie naine	Ansar Chico	Zwerggans	Ganso-pequeno-detesta-branca
II	*Anser fabalis*	Bean Goose	Oie des moissons	Ansar Campestre	Saatgans	Ganso-campestre, Ganso-de-bico-curto
I/II	*Aquila adalberti* Formerly included in *Aquila heliaca* (s.l.)	Adalbert's Eagle	Aigle ibérique	Aguila Imperial Ibérica	Spanischer Kaiseradler	Águia-imperial-ibérica
I/II	*Aquila clanga*	Greater Spotted Eagle	Aigle criard	Aguila Moteada	Schelladler	Águia-gritadeira
I/II	*Aquila heliaca*	Imperial Eagle	Aigle impérial	Aguila Imperial Oriental	Kaiseradler	Águia-imperial-oriental
II	*Arctocephalus australis*	South American Fur seal	Otarie d'Amérique du Sud	Lobo fino Sudamericano, Lobo fino austral	Südliche Pelzrobbe	Lobo-marinho-sul-americano
II	*Ardea purpurea purpurea*	Purple Heron	Héron pourpré	Garza Imperial	Purpurreiher	Garça-vermelha
I/II	*Ardeola idae*	Malagasy Pond Heron	Crabier blanc	Garcilla Malgache	Rotbauchreiher	Garça-caranguejeira de Madagascar
II	*Ardeola rufiventris*	Rufous-bellied Heron	Héron à ventre roux	Garcilla Ventriroja	Dickschnabel reiher	Garça-de-barrigavermelha
II	*Arenaria interpres*	Ruddy Turnstone	Tournepierre à collier	Vuelvepiedras Común	Steinwälzer	Rola-do-mar, Virapedras
I	*Aythya baeri*	Baer's pochard	Fuligule de Baer	Porrón de Baer	Baerente, Schwarzkopf-moorente	Zarro de Baer
II	*Aythya ferina*	Common Pochard	Fuligule milouin	Porrón Europeo	Tafelente	Zarro-comum
II	*Aythya fuligula*	Tufted Duck	Fuligule morillon	Porrón Moñudo	Reiherente	Zarro-negrinha
II	*Aythya marila*	Greater Scaup	Fuligule milouinan	Porrón Bastardo	Bergente	Zarro-bastardo
I/II	*Aythya nyroca*	Ferruginous Pochard, Ferruginous Duck	Fuligule nyroca	Porrón Pardo	Moorente	Zarro-castanho
I	*Balaena mysticetus*	Bowhead Whale	Baleine du Groenland, Baleine franche	Ballena de Groenlandia	Grönlandwal	Baleia-da-groenlândia
II	*Balaenoptera bonaerensis*	Antartic Minke whale	Petite rorqual de l'Antarctique	Rorcual enano del antarctica	Südliche Zwergwal	Baleia-de-Minke, Baleia-minke-antártica
I/II	*Balaenoptera borealis*	Sei Whale, Coalfish whale, Pollack whale, Rudolph's Rorqual	Rorqual Sei, Baleinoptère de Rudolphi, Rorqual boréal, Rorqual de Rudolphi	Ballena Sei, Ballena Boba, Rorcual Boreal, Rorcual de Rudolphi, Rorcual Norteno	Seiwal	Baleia sei, Baleiaboreal, Baleia-glacial, Baleia-sardinheira

App	Taxon	English	Français	Español	Deutsch	Português
II	*Balaenoptera edeni*	Bryde's whale, Tropical whale	Baleinoptère de Bryde, Rorqual d'Eden, Rorqual de Bryde, Rorqual tropical	Ballena de Bryde	Edenwal, Brydewal!	Baleia-de-bryde
I	*Balaenoptera musculus*	Blue Whale	Baleine bleue, grand Rorqual	Ballena Azul	Blauwal	Baleia-azul
II	*Balaenoptera omurai Formerly included in Balaenoptera edeni*					
I/II	*Balaenoptera physalus*	Fin Whale	Baleine fin, Baleine à Nageoires, Baleinoptère commun, Rorqual commun	Ballena Aleta, Ballena Boba, Rorcual Común	Finnwal	Baleia-comum, Baleiafin, Rorqual-comum
II	*Barbastella barbastellus*	Barbastelle Bat	Barbastelle d'Europe	Murciélago de Bosque	Mopsfledermaus	Morcego-negro
II	*Berardius bairdii*	Baird's Beaked Whale	Baleine à bec de Baird	Zifio de Baird	Baird-Schnabelwal	Baleia-bicuda-de-baird
I	*Bos grunniens*	Wild Yak, Yak	Yack sauvage	Yak	Wildyak	Iaque
I	*Bos sauveli*	Kouprey	Kouprey	Kouprey	Kuprey	Kouprey
II	*Botaurus stellaris stellaris*	Eurasian Bittern	Butor étoilé	Avetoro Común	Rohrdommel	Abetouro-comum
II	*Branta bernicla*	Brent Goose	Bernache cravant	Barnacla Carinegra	Ringelgans	Ganso-de-faces-negras
II	*Branta leucopsis*	Barnacle Goose	Bernache nonnette	Barnacla Cariblanca	Nonnengans	Ganso-de-faces-brancas
I/II	*Branta ruficollis*	Red-breasted Goose	Bernache à cou roux	Barnacla Cuelliroja	Rothalsgans	Ganso-de-pescoço-ruivo
I	*Brotogeris pyrrhopterus*	Grey-cheeked Parakeet	Toui flamboyant	Perico macareño	Feuerflügelsittich	Periquito-de-bochechacinza
II	*Bucephala clangula*	Common Goldeneye	Garrot à oeil d'or	Porrón Osculado	Schellente	Pato-ouro-d'ouro
II	*Burhinus oedicnemus*	Stone Curlew	Oedicnème criard	Alcaraván	Triel	Alcavarão
II	*Calidris alba*	Sanderling	Bécasseau sanderling	Correlimos Tridáctilo	Sanderling	Maçarico-branco, Pilrito-das-praias
II	*Calidris alpina*	Dunlin	Bécasseau variable	Correlimos Común	Alpenstrandläufer	Pilrito-comum
I/II	*Calidris canutus rufa*	Red Knot	Bécasseau maubèche	Correlimos Gordo	Knutt	Ruiva
II	*Calidris ferruginea*	Curlew Sandpiper	Bécasseau cocorli	Correlimos Zarapitín	Sichelstrandläufer	Pilrito-de-bicocomprido
II	*Calidris maritima*	Purple Sandpiper	Bécasseau violet	Correlimos Oscuro	Meerstrandläufer	Pilrito-escuro
II	*Calidris minuta*	Little Stint	Bécasseau minute	Correlimos Menudo	Zwergstrandläufer	Pilrito-pequeno
II	*Calidris temminckii*	Temminck's Stint	Bécasseau de Temminck	Correlimos de Temminck	Temminck-strandläufer	Pilrito-de-Temminck
II	*Calidris tenuirostris*	Great Knot	Grand Bécasseau maubèche, Bécasseau de l'Anadyr	Correlimos Grande	Anadyr-Knutt, Großer Knutt	Pilro de Bering
I	*Camelus bactrianus*	Wild or Bactrian camel	Chameau de Bactriane	Camello Bactriano	Wildkamel	CameloBactriano
II	*Caperea marginata*	Pygmy Right whale	Baleine pygmée	Ballena franca pigmea	Zwergglattwal	Baleia-franca-pigméia
I/II	*Carcharodon carcharias*	Great White Shark, White Shark	Grand requin blanc, le grand requin	Jaquetón blanco, Marraco, Gran tiburón blanco	Weißer Hai	Tubarão-branco
I/II	*Caretta caretta*	Loggerhead Turtle	Caouanne, Tortue carette	Tortuga Boba	Unechte Karettschildkröte	Tartaruga-comum, Tartaruga-amarela, cabeçuda, Tartaruga-meiopente, Tartarugamestiça
II	*Casmerodius albus albus*	Great Egret, Great White Egret	Grande aigrette	Garceta Grande	Silberreiher	Garça-branca-grande
II	*Cephalorhynchus*	Commerson's Dolphin	Dauphin de Commerson	Delfín de Commerson	Commerson-Delphin	Golfinho-de-Commerson

App	Taxon	English	Français	Español	Deutsch	Português
II	*Cephalorhynchus eutropia*	Chilean Dolphin	Dauphin du Chili	Delfín Chileno	Weißbauchdelfin, Chilenischer Delfin	Golfinho-chileno
II	*Cephalorhynchus heavisidii*	Heaviside's Dolphin	Céphalorhynque du Cap	Delfín del Cabo	Heaviside-Delphin	Golfinho-de-heaviside
I/II	*Cervus elaphus yarkandensis* Formerly listed as *Cervus elaphus bactrianus*	Bukhara Deer	Cerf Bactrian; Cerf Bukharian	Ciervo de Berbería (EU CITES)	Baktrischer Rothirsch; Bucharahirsch	Veado-de-yarkand
I	*Cervus elaphus barbarus*	Barbary Stag, Barbary Deer	Cerf de Barbarie	Ciervo de Berbería Ciervo de Cachemira (EU CITES)	Berberhirsch	Veado-da-Bárbaro, Veado-berbere
I/II	*Cetorhinus maximus*	Basking shark	Requin pèlerin	Tiburón peregrine	Riesenhai	Tubarão-elefante
II	*Charadrius alexandrinus*	Kentish Plover	Gravelot à collier interrompu, Pluvier à collier interrompu	Chorlitejo Patinegro	Seeregenpfeifer	Borrelho-de-colheira-interrompida
II	*Charadrius asiaticus*	Caspian Plover	Pluvier asiatique	Chorlitejo Asiático Chico	Wermutregen-pfeifer	Borrelho-asiatico
II	*Charadrius dubius*	Little Ringed Plover	Petit gravelot, Pluvier petit-gravelot	Chorlitejo Chico	Flussregenpfeifer	Borrelho-pequeno-decoleira
II	*Charadrius forbesi*	Forbes' Plover	Pluvier de Forbes	Chorlitejo de Forbes	Braunstirnregen-pfeifer, Forbesregenpfeifer	Borrelho-de-forbe
II	*Charadrius hiaticula*	Common Ringed Plover	Grand gravelot, Pluvier grand-gravelot	Chorlitejo Grande	Sandregenpfeifer	Borrelho-grande-decoleira
II	*Charadrius leschenaultii*	Greater Sandplover	Pluvier du désert, Pluvier de Leschenault	Chorlitejo Mongol Grande	Wüstenregen-pfeifer	Borrelho-mongolgrande
II	*Charadrius marginatus*	White-fronted Plover	Pluvier à front blanc	Chorlitejo Frentiblanco pfeifer	Weißstirnregen-	Borrelho-de-testabranca
II	*Charadrius mongolus*	Mongolian Plover, Lesser Sandplover	Gravelot de Mongolie, Pluvier de Mongolie	Chorlitejo Mongol Chico	Mongolenregen-pfeifer	Borrelho-mongol
II	*Charadrius pallidus*	Chestnut-banded Plover	Pluvier élégant	Chorlitejo Pálido	Rotbandregen-pfeifer	Borrelho-pálido, Borrelho-de-colararruivado
II	*Charadrius pecuarius*	Kittlitz's Plover	Gravelot pâtre, Pluvier pâtre	Chorlitejo Pecuario	Hirtenregenpfeifer	Borrelho-do-gado, Borrelho de Kittli
II	*Charadrius tricollaris*	Three-banded Plover	Pluvier à triple collier	Chorlitejo Tricollar	Dreibandregen-pfeifer	Borrelho-de-triplacoleira, Borrelho-detrês-golas
I/II	*Chelonia mydas*	Green Turtle	Tortue verte	Tortuga Verde	Pazifische Suppenschildkröte	Tartaruga-verde, Aruanã
I/II	*Chlamydotis undulata*	Houbara Bustard	Outarde Houbara	Hubara	Kragentrappe	Abetarda-moura
II	*Chlidonias leucopterus*	White-winged Black	Guifette leucoptère	Fumarel Aliblanco	Wießflügelsee-schwalbe	Gavina-d'asa-branca, Trinta-reis-negro
II	*Chlidonias niger niger*	Black Tern	Guifette noire	Fumarel Común	Trauerseeschwalbe	Gavina-preta
I/II	*Chloephaga rubidiceps*	Ruddy-headed Goose	Oie des andes à tête rousse	Cauquén Colorado	Rotkopfgans	
I	*Ciconia boyciana*	Oriental White Stork	Cigogne à bec noir, Cigogne blanche du Japon	Cigüeña Oriental	Schwarzschnabel storch	Cegonha brancaoriental
II	*Ciconia ciconia*	White Stork	Cigogne blanche	Cigüeña Blanca	Weißstorch	Ceginha-brance
II	*Ciconia episcopus microscelis*	Woolly-necked Stork	Cigogne épiscopale	Cigüeña Lanuda	Afrikanischer Wollhalsstorch	Cegonha-episcopal

App	Taxon	English	Français	Español	Deutsch	Português
II	*Ciconia nigra*	Black Stork	Cigogne noire	Cigüeña Negra	Schwarzstorch	Cegonha-preta
II	*Clangula hyemalis*	Long-tailed Duck	Harelde de Miquelon, Harelde Kakawi	Pato Havelda	Eisente	Pato-de-cauda-afilada
II	*Coracias garrulus*	Roller	Rollier d'Europe	Carraca	Blauracke	Roliero-comum, Roliero-europeu
II	*Coturnix coturnix coturnix*	Quail	Caille des blés	Codorniz Común	Wachtel	Codorna
II	*Crex crex*	Corncrake	Râle des genêts	Guión de Codornices	Wachtelkönig	Codornizão
II	*Crocodylus porosus*	Salt-water Crocodile, Estuarine Crocodile	Crocodile marin	Cocodrilo Poroso águasalgada	Leistenkrokodil	Crocodilo-de-
II	*Cygnus columbianus*	Bewick's Swan	Cygne de Bewick	Cisne Chico	Zwergschwan	Cisne-pequeno
II	*Cygnus cygnus*	Whooper Swan	Cygne chanteur	Cisne Cantor	Singschwan	Cisne-bravo
II	*Cygnus olor*	Mute Swan	Cygne tuberculé	Cisne Vulgar	Höckerschwan	Cisne-branco, Cisnevulgar
II	*Danaus plexippus*	Monarch Butterfly	Papillon monarque	Mariposa Monarca	Monarchfalter	Borboleta-monarca
II	*Delphinapterus leucas*	White Whale, Beluga	Belouga, Dauphin blanc Marsouin	Beluga, Ballena Blanca	Weißwal, Beluga	Baleia-branca, Beluga
II	*Delphinus delphis*	Common Dolphin	Dauphin commun	Delfín Común	Gemeiner Delphin	Golfinho-comum-debico-curto
II	*Dendrocygna bicolor*	Fulvous Whistling-duck	Dendrocygne fauve	Suirirí Bicolor, Pato Silbón, Pato Silbón Rojizo	Gelbbrust-Pfeifgans	Marreca-caneleira, Marreca-peba
II	*Dendrocygna viduata*	White-faced Whistling-duck	Dendrocygne veuf	Suirirí Cariblanco, Pato Cara Blanca	Witwenpfeifgans	Irerê
I	*Dendroica caerulea*	Cerulean warbler	Fauvette azurée; Paruline azurée	Bijirita azulosa; Verdín azulado; Gorjeador ceruleo; Chipe ceruleo; Reinita cerulea	Pappelwaldsänger	Mariquitaazul
I	*Dendroica kirtlandii*	Kirtland's Warbler	Figuier de Kirtland	Silbador de Kirtland	Kirtlands Waldsänger, Michiganwalds-änger	
I/II	*Dermochelys coriacea*	Leatherback Turtle, Leathery Turtle	Tortue luth	Tortuga Laúd	Lederschildkröte	Tartaruga-de-couro
I	*Diomedea albatrus*	Short-tailed Albatross, Steller's Albatross	Albatros à queue courte, Albatros de Steller	Albatros Colicorto	Kurzschwanz-albatros	Albatroz-de-cauda-curta
I	*Diomedea amsterdamensis*	Amsterdam Albatross	Albatros d'Amsterdam	Albatros de la Amsterdam	Amsterdam -Albatros	Albatroz-de-amsterdam
II	*Diomedea bulleri*	Buller's Albatross	Albatros de Buller	Albatros de Buller	Bulleralbatros	Albatroz-de-buller
II	*Diomedea cauta*	Shy Albatross	Albatros à cape blanche	Albatros Frentiblanco	Weißkappen-albatros	Albatroz-arisco
II	*Diomedea chlororhynchos*	Yellow-nosed Albatross	Albatros à bec jaune	Albatros Clororrinco, Albatros de Pico Fino	Gelbnasen-albatros	Albatroz-de-bicoamarelo
II	*Diomedea chrysostoma*	Grey-headed Albatross	Albatros à tête grise	Albatros Cabecigris, Albatros de Cabeza Gris	Graukopf-albatros	Albatroz-de-cabeçacinza
II	*Diomedea epomophora*	Royal Albatross	Albatros royal	Albatros Real	Königsalbatros	Albatroz-real
II	*Diomedea exulans*	Wandering Albatross	Albatros huleur	Albatros Viajero	Wanderalbatros	Albatroz-gigante, Albatroz-errante
II	*Diomedea immutabilis*	Laysan Albatross	Laysanalbatros	Albatros de Laysan	Laysan-Albatros	Albatroz-de-laysan
II	*Diomedea irrorata*	Waved Albatross	Albatros des Galapagos	Albatros de las Galapagos	Galapagos-Albatros	Albatroz-das-galápagos
II	*Diomedea melanophris*	Black-browed Albatross	Albatros à sourcils noirs	Albatros Ojeroso, Albatros de Ceja Negra	Mollymauk	Albatroz-desobrancelha

App	Taxon	English	Français	Español	Deutsch	Português
II	*Diomedea nigripes*	Black-footed Albatross	Albatros à pieds noirs	Albatros Patinegro	Schwarzfuß-albatros	Albatroz-patinegro, Albatroz-dos-péspretos
II	*Dromas ardeola*	Crab Plover	Drome ardéole	Cigüeñela Cangrejera	Reiherläufer	Caranguejeiro
II	*Dugong dugon*	Dugong, Sea Cow	Dugong	Dugongo	Dugong, Pazifische Seekuh	Dugongo
I	*Egretta eulophotes*	Chinese Egret	Aigrette de Chine	Garceta China	Schneereiher, Chinaseidenreiher	
II	*Egretta vinaceigula*	Slaty Egret	Aigrette vineuse	Garceta Gorgirroja	Braunkehlreiher	Garça-de-gargantavermelha
II	*Eidolon helvum*	Straw-coloured fruit bat	Roussette des palmiers africaine		Palmenflughund	Morcego-frugívoro
I	*Emberiza aureola*	Yellow-breasted bunting	Bruant auréole	Escribano aureolado	Weidenammer	Escrevedeira-aureolada
II	*Eptesicus nilssonii*	Northern Serotine Bat	Sérotine boréale	Murciélago Norteño	Nordfledermaus	Morcego -hortelão-donorte
II	*Eptesicus serotinus*	Serotine Bat	Grande sérotine	Murciélago Hortelano	Breitflügelf-ledermaus	Morcego-hortelão
I	*Equus grevyi*	Grevy's Zebra	Zèbre de Grevy	Cebra de Grevy	Grevyzebra	Zebra-de-grevy
II	*Equus hemionus* Includes *Equus onager*	Asiatic wild ass	Ane sauvage de l'Asie	Asno salvaje asiatico	Asiatischer Wildesel	Asno, Hemíono
II	*Equus kiang* Formerly included in *Equus hemionus* (s.l.)	Kiang	Kiang	Kiang	Kiang	Kiang
I/II	*Eretmochelys imbricata*	Hawksbill Turtle	Tortue imbriquée, Caret	Tortuga Carey	Echte Karettschildkröte	Tartaruga-de-pente
I	*Eubalaena australis* Formerly listed as *Balaena glacialis australis*	Southern Right Whale	Baleine australe	Ballena Franca Austral	Südkaper, südlicher Glattwal	Baleia-franca, Baleiafranca-austral
I	*Eubalaena glacialis* Formerly listed as *Balaena glacialis glacialis*	Northern Right Whale, Biscayan Right Whale	Baleine de Biscaye	Ballena Franca	Nordkaper	Baleia-franca-do-atlântico-norte
I	*Eubalaena japonica* Formerly listed as *Balaena glacialis glacialis*	North Pacific Right Whale	Baleine (franche) du Pacifique Nord	Ballena franca del Pacífico norte	Pazifischer Nordkaper	Baleia-franca-dopacífico
II	*Eudromias morinellus*	Eurasian Dotterel	Pluvier guignard	Chorlito Carambolo	Mornellregen-pfeifer	Borrelho-ruivo, Tarambola-carambola
I/II	*Eurynorhynchus pygmeus*	Spoon-billed Sandpiper	Bécasseau spatule	Correlimos cuchareta	Loeffelstranlaeufer	
I/II	*Falco naumanni*	Lesser Kestrel	Faucon crécerellete	Cernícalo Primilla	Rötelfalke	Peneireiro-das-torres
II	*Fulica atra atra*	Common Coot	Foulque macroule	Focha Común	Blässhuhn	Galeirão-comum
II	*Gallinago gallinago*	Common Snipe	Bécassine des marais	Agachidiza Común	Bekassine	Narceja-comum
II	*Gallinago media*	Great Snipe, Double Snipe	Bécassine double	Agachadiza Real	Doppelschnepfe	Narceja-real
II	*Gavia adamsii*	White-billed Diver	Plongeon à bec blanc	Colimbo de Adams	Gelbschnabel-Eistaucher	Mobêlha-de-bicobranco
II	*Gavia arctica arctica*	Black-throated Diver	Plongeon arctique	Colimbo Artico	Prachttaucher	Mobêlha-ártica
II	*Gavia arctica suschkini*	Black-throated Diver	Plongeon arctique	Colimbo Artico	Prachttaucher	Mobêlha-ártica
II	*Gavia immer immer*	Great Northern Diver	Plongeon imbrin	Colimbo Grande	Eistaucher	Mobêlha-grande
II	*Gavia stellata*	Red-throated Diver	Plongeon catmarin	Colimbo Chico	Sterntaucher	Mobêlha-pequena
I	*Gavialis gangeticus*	Gharial, Indian Gavial	Gavial du Gange	Gavial del Ganges	Ganges-Gavial	Gavial

App	Taxon	English	Français	Español	Deutsch	Português
I	*Gazella cuvieri*	Cuvier's Gazelle	Gazelle de Cuvier	Gacela de Cuvier	Afrikanische Echtgazelle, Atlasgazelle	Gazela-de-cuvier
I	*Gazella dorcas*	Dorcas Gazelle	Gazelle dorcas	Gacela Dorcas	Dorkasgazelle	Gazela-dorcas
II	*Gazella erlangeri* Formerly included in *Gazella gazella*	Neumann's Gazelle				
II	*Gazella gazella*	Mountain Gazelle, Edmi Gazelle	Gazelle de Edmi, Gazelle d'Arabie	Gacela de la India	Edmigazelle	Gazela-de-Edmi
I	*Gazella leptoceros*	Slender-horned Gazelle, Rhim	Gazelle leptocère	Gacela de Astas Delgadas	Dünengazelle	Gazela-de-chifresbrancos
II	*Gazella subgutturosa*	Goitered or Black-tailed gazelle	Gazelle à goitre	Gacela persa	Kropfgazelle, persische Gazelle	Gazela-persa
I/II	*Geronticus eremita*	Waldrapp, Hermit	Ibis Ibis chauve, Comatibis chevelu	Ibis Eremita	Waldrapp	Íbis-calva, Íbis-eremita
II	*Glareola nordmanni*	Black-winged Pratincole	Glaréole à ailes noires	Canastera Alinegra	Schwarzflügel-brachschwalbe	Perdiz-do-mar-d'asapreta
II	*Glareola nuchalis*	Rock Pranticole	Glaréole auréolée	Canastera Sombría	Halsband-	Perdiz-do-mar-escura Brachschwalbe
II	*Glareola pratincola*	Collared Pratincole	Glaréole à collier	Canastera	Rotflügelbrach-schwalbe	Perdiz-do-mar
II	*Globicephala melas*	Long-finned Pilot Whale	Globicéphale noir, Chandron	Calderón Negro	Grindwal	Baleia-piloto-de-aletalonga, Baleia-piloto-depeitorais-longas
I	*Gorilla beringei* Formerly included in *Gorilla gorilla*	Eastern Gorilla	Gorille de l'est	Gorila oriental	Östlicher Gorilla	Gorila-do-oriente
I	*Gorilla gorilla*	Western Gorilla	Gorille de l'Ouest	Gorila occidental	Westlicher Gorilla	Gorila-do-ocidente
I	*Gorsachius goisagi*	Japanese Night Heron	Bihoreau goisagi	Bihoreau goisagi	Rotschnabelreiher	Garça-noturna-japonesa
II	*Grampus griseus*	Risso's Dolphin	Dauphin de Risso, Marsouin gris	Delfín de Risso, Calderón gris	Rundkopf-delphin	Golfinho-de-risso
II	*Grus carunculatus*	Wattled Crane	Grue caronculée	Grulla Carunculada	Klunkerkranich	Grou-carunculado
II	*Grus grus*	Common Crane	Grue cendrée	Grulla Común	Graukranich	Grou-comum
I/II	*Grus japonensis*	Manchurian Crane, Japanese Crane	Grue du Japon, Grue de Manchourie	Grulla de Manchuria	Mandschuren-kranich, Japankranich	Grou-japonês, Grouda manchúria, Grou-decrista-vermelha
I/II	*Grus leucogeranus*	Siberian Crane	Grue de Sibérie, Grue blanche	Grulla Siberiana, Grulla Blanca	Schneekranich, Nonnenkranich	Grou-Branco
I/II	*Grus monacha*	Hooded Crane	Grue moine	Grulla monjita	Mönchskranich	Grou-de-capuz
I/II	*Grus nigricollis*	Black-necked Crane	Grue à cou noir	Grulla Cuellinegra	Schwarzhals-kranich	Grou-de-capuz-preto
II	*Grus paradisea*	Blue Crane	Grue de paradis	Grulla del Paraíso	Paradieskranich	Grou-do-paraíso
I/II	*Grus vipio*	White-naped Crane	Grue à cou blanc	Grulla Cuelliblanca	Weißnacken-kranich	Grou-de-pescoçobranco
II	*Grus virgo* Formerly listed as *Anthropoides virgo*	Demoiselle Crane	Grue demoiselle	Grulla Damisela	Jungfernkranich	Grou-pequeno

App	Taxon	English	Français	Español	Deutsch	Português
I/II	*Haliaeetus albicilla*	White-tailed Eagle	Pygargue à queue blanche, Pygargue commun	Pigargo Europeo	Seeadler	Águiarabalva
I/II	*Haliaeetus leucoryphus*	Pallas' Sea-Eagle, Pallas' Fishing Eagle	Pygargue de Pallas	Pigargo de Pallas	Bandseeadler, Seeadler?, Fischadler?	Águia de Pallas
I/II	*Haliaeetus pelagicus*	Steller's Sea Eagle	Pygargue de Steller	Pigargo Gigante	Riesenseeadler	Águia-do-mar-de-steller
II	*Halichoerus grypus*	Grey Seal	Phoque gris	Foca Gris	Kegelrobbe	Foca-cinzenta
II	*Himantopus himantopus*	Black-winged Stilt	Echasse blanche	Avocetas, Cigüeñuela Común	Stelzenläufer	Pernilongo, Pernilongode-costas-negras
I	*Hippocamelus bisulcus*	South Andean Deer	Cerf des Andes méridionales	Huemul	Südandenhirsch, Südlicher Andenhirsch, Huemul	Cervo-sul-andino, Huemul
I/II	*Hirundo atrocaerulea*	Blue Swallow	Hirondelle bleue	Golondrina azul	Stahlschwalbe	Andorinha-Azul
II	*Huso dauricus*	Kaluga Sturgeon	Esturgeon Kaluga	Esturión kaluga	Kaluga-Hausen	Esturjão-kaluga
II	*Huso huso*	Giant Sturgeon, Beluga	Beluga	Beluga, Esturión gigante	Hausen	Esturjão-branco, Esturjão-beluga
II	*Hyperoodon ampullatus*	Northern Bottlenose Whale, Bottle-nosed Whale	Hyperoodon boréal	Ballena Morro de Botella, Ballena Nariz de Botella	Dögling, Entenwal	Bakeia-de-nariz-degarrafa, Botinhoso, Baleias-de-bico-degarrafa
II	*Inia geoffrensis*	Amazon River Dolphin, Boutu	Dauphin de l'Amazone	Delfín Rosado del Amazonas	Amazonas-Delphin	Boto-cor-de-rosa
II	*Isurus oxyrinchus*	Shortfin Mako shark	Requin-Taupe bleu	Marrajo dientuso	Kurzflossen-Mako	Tubarão-mako, Tubarão-mako-cavala
II	*Isurus paucus*	Longfin Mako shark	Requin petite-taupe	Marrajo carite	Langflossen-Mako	Anequim-preto
II	*Ixobrychus minutus minutus*	Little Bittern	Blongios nain Avetorillo	Avetorillo Común	Zwergdommel	Garça-pequena
II	*Ixobrychus sturmii*	African Dwarf Bittern	Blongios de Sturm	Avetorillo Plomizo	Schieferdommel	Garçote-anão
II	*Lagenodelphis hosei*	Fraser's Dolphin	Dauphin de Fraser	Delfín de Fraser	Borneo-Delphin	Golfinho-de-fraser
II	*Lagenorhynchus acutus*	Atlantic White-sided Dolphin	Dauphin à flancs blancs	Delfín de Costados Blancos	Weißseiten-delphin	Golfinho-de-lateraisbrancas
II	*Lagenorhynchus albirostris*	White-beaked Dolphin	Dauphin à bec blanc, Dauphin à rostre blanc	Delfín de Pico Blanco	Weißschnauzen-delphin	Golfinho-de-bicobranco
II	*Lagenorhynchus australis*	Peale's Dolphin, Blackchin Dolphin	Dauphin de Peale	Delfín Austral	Peale Delphin	Golfinho-de-peale, Golfinho-do-sul
II	*Lagenorhynchus obscurus*	Dusky Dolphin	Dauphin obscur	Delfín Oscuro	Schwarzdelphin	Golfinho-escuro
II	*Lamna nasus*	Porbeagle	Requin-taupe commun	Marrajo sardinero; cailón marrajo, moka, pinocho	Heringshai	Marracho
II	*Larus armenicus*	Armenian Gull	Goéland d'Armenie	Gaviota Armenia	Armenienmöwe	Gaivota-da-arménia
I	*Larus atlanticus*	Olrog's Gull	Goéland d'Olrog	Gaviota de Olrog, Gaviota Cangrejera	Olrogmöwe	Gaivota-de-rabo-preto
I/II	*Larus audouinii*	Audouin's Gull	Goéland d'Audouin	Gaviota de Audouin	Korallenmöwe	Gaivota-de-adouin
II	*Larus genei*	Slender-billed Gull	Goéland railleur	Gaviota Picofina	Dünnschnabel-möwe	Gaivota-de-bico-fino
II	*Larus hemprichii*	Sooty Gull, Hemprich's Gull, Aden Gull	Goéland d'Hemprich	Gaviota de Adén Piquiverde, Gaviota Cejiblanca	Hemprichmöwe	Gaivota-fuliginosa

App	Taxon	English	Français	Español	Deutsch	Português
II	*Larus ichthyaetus*	Great Black-headed Gull	Goéland ichthyaète	Gavión Cabecinegro	Fischmöwe	Alcatraz-de-cabeçapreta
I/II	*Larus leucophthalmus*	White-eyed Gull	Goéland à iris blanc, Goéland à collier blanc	Gaviota Ojiblanca, Gaviota de Adén Piquirroja	Weißaugen-möwe	Gaivota-d'olho-branco
II	*Larus melanocephalus*	Mediterranean Gull	Mouette mélanocéphale	Gaviota Cabecinegra	Schwarzkopf-möwe	Gaivota-de-cabeçapreta, Gaivota-domediterrâneo
I	*Larus relictus*	Relict Gull	Goéland de Mongolie, Goéland relique	Gaviota de Mongolia	Gobi-Schwarz-koptmöwe, Lönnbergmöwe	Gaivota da Mongólia
I	*Larus saundersi*	Saunder's Gull, Chinese Black-headed Gull	Goéland de Saunders	Gaviota de Saunders	Kappenmöwe, Saundersmöwe	Gaivota-de-saunders
I/II	*Lepidochelys kempii*	Kemp's Ridley Turtle, Atlantic Ridley Turtle	Tortue de Ridley, Caret des Antilles	Tortuga Lora	Atlantische Bastardschildkröte	Tartaruga-de-kemp
I/II	*Lepidochelys olivacea*	Ridley Turtle, Olive Ridley Turtle	Tortue bâtarde	Tortuga Olivácea	Bastardschildkröte	Tartaruga-oliva, Tartaruga-olivácea
II	*Limicola falcinellus*	Broad-billed Sandpiper	Bécasseau falcinelle	Correlimos Falcinelo	Sumpfläufer	Pilrito-falcinelo, Maçarico-branco
II	*Limosa lapponica*	Bar-tailed Godwit	Barge rousse	Aguja Colipinta	Pfuhlschnepfe	Fuselo
II	*Limosa limosa*	Black-tailed Godwit	Barge à queue noire	Aguja Colinegra	Uferschnepfe	Maçarico-de-bicodireito
I	*Lontra felina* Formerly listed as *Lutra felina*	Marine Otter	Loutre de Mer	Chungungo	Meerotter	Chuchungo
I	*Lontra provocax* Formerly listed as *Lutra provocax*	Southern River Otter	Loutre du Chili	Huillín	Südlicher Flussotter	Huillín, Lontra da Argentina
II	*Loxodonta africana*	African Elephant	Eléphant d'Afrique	Elefante Africano	Afrikanischer Elefant	Elefante-africano
II	*Loxodonta cyclotis* Formerly included in *Loxodonta africana*	African forest elephant	Eléphant de forêt d'Afrique	Elefante africano de bosque	Waldelefant	Elefante-africano-desavana
II	*Lycaon pictus*	African wild dog	Lycaon	Licaon	Afrikanischer Wildhund	Cachorro-selvagemafricano
II	*Lymnocryptes minimus*	Jack Snipe	Bécassine sourde	Agachadiza Chica	Zwergschnepfe	Narceja-galega
II	*Macronectes giganteus*	Southern Giant Petrel	Fulmar géant	Abanto-marino Antártico, Petrel Gigante Común	Riesensturmvogel	Petrel-gigante
II	*Macronectes halli*	Northern Giant Petrel	Fulmar de Hall	Abanto-marino Subantártico, Petrel Gigante	Hallsturmvogel	Petrel-gigante-do-norte
I/II	*Marmaronetta angustirostris*	Marbled Teal	Sarcelle marbrée, Marmaronette marbrée	Cerceta Pardilla	Marmelente	Pardilheira
I	*Megaptera novaeangliae*	Humpback Whale	Mégaptère	Yubarta	Buckelwal	Baleia-jubarte
II	*Melanitta fusca*	Velvet Scoter, Whitewinged Scoter	Macreuse brune	Negrón Especulado	Samtente	Pato-fusco
II	*Melanitta nigra*	Common Scoter, Black Scoter	Macreuse noire	Negrón Común	Trauerente	Pato-preto
II	*Mergellus albellus*	Smew	Harle piette	Serreta Chica	Zwergsäger	Merganso-pequeno
II	*Mergus merganser*	Goosander, Common Merganser	Harle bièvre, Grand harle	Serreta Grande	Gänsesäger	Marganso-grande

App	Taxon	English	Français	Español	Deutsch	Português
II	*Mergus serrator*	Red-breasted Merganser	Harle huppé	Serreta Mediana	Mittelsäger	Merganso-de-poupa
II	*Merops apiaster*	Bee-eater	Guêpier	Abejaruco	Bienenfresser	Abelharuco-comum
II	*Miniopterus majori* Formerly included in *Miniopterus schreibersii*					
II	*Miniopterus natalensis* Formerly included in *Miniopterus schreibersii*	Natal long-fingered Bat				Morcego-de-natal-dededos-longos
II	*Miniopterus schreibersii*	Schreiber's Bent -winged Bat	Minoptère à longues ailes	Murciélago de Cueva	Langflügel-fledermaus	Morcego-de-peluche
I/II	*Monachus monachus*	Mediterranean Monk Seal	Phoque moine de Méditerranée	Foca Monje del Mediterráneo	Mittelmeer-Mönchsrobbe	Foca-monge-domediterrâneo
II	*Monodon monoceros*	Narwhal	Narval	Narval	Narwal	Narval
II	*Mycteria ibis*	Yellow-billed Stork	Tantale ibis	Tántalo Africano	Nimmersatt	Cegonha-de-bicoamarelo
II	*Myotis bechsteini*	Bechstein's Bat	Vespertilion de Bechstein	Murciélago de Bechstein	Bechstein-Fledermaus	Morcego-de-bechstein
II	*Myotis blythi*	Lesser Mouse-eared Bat	Petit murin	Ratonero Mediano	Kleines Mausohr	Morcego-rato-pequeno
II	*Myotis brandtii*	Brandt's Bat	Vespertilion de Brandt	Murciélago de Brandt	Großbart-fledermaus	
II	*Myotis capaccinii*	Long-fingered Bat	Vespertilion de Capaccini	Murciélago Patudo	Langfußfledermaus	
II	*Myotis dasycneme*	Pond Bat	Vespertilion des marais	Murciélago Lagunero	Teichfledermaus	
II	*Myotis daubentoni*	Daubenton's Bat	Vespertilion de Daubenton	Murciélago Ribereño	Wasserfledermaus	Morcego-de-agua
II	*Myotis emarginatus*	Geoffroy's Bat, Notcheared Bat	Vespertilion à oreilles échancrées	Murciélago de Geoffroy	Wimperfledermaus	Morgeco-lanudo
II	*Myotis myotis*	Greater Mouse-eared Bat	Grand murin	Murciélago Ratonero Grande	Großes Mausohr	Morcego-rato-grande
II	*Myotis mystacinus*	Whiskered Bat	Vespertilion à moustaches	Murciélago Bigotudo	Kleine Bart-fledermaus	Morcego-de-bigodes
II	Myotis nattereri	Natterer's Bat	Vespertilion de Natterer	Murciélago de Natterer	Fransenfledermaus	Morcego-de-franja
I	*Nanger dama* Formerly listed as *Gazella dama*	Dama Gazelle	Gazelle dama	Gacela Dama	Damagazelle	Gazella-dama
II	*Natator depressus*	Flatback Turtle	Tortue à dos plat	Tortuga Kikila	Australische Suppenschild-kröte	Tartaruga-de-caso-achatado, Tartaruga-australiana
II	*Neophocaena phocaenoides*	Finless Porpoise	Marsouin noir, Marsouin de l'Inde	Marsopa Negra	Indischer Schweinswal	Boto-do-índico
I	*Neophron percnopterus*	Egyptian vulture	Vautour percnoptère	Alimoche común	Schmutzgeier	Abutre-do-egito
II	*Netta erythrophthalma*	Southern Pochard	Nette brune	Pato Morado	Rotaugenente	Paturi-preta
II	*Netta rufina*	Red-crested Pochard	Nette rousse	Pato Colorado	Kolbenente	Pato-de-bico-vermelho
II	*Nettapus auritus*	African Pygmy-goose	Sarcelle à oreillons, Anserelle naine	Gansito Africano	Afrikanische Zwergente, Rostbrust-Zwerggans	Pato-orelhudo
II	*Numenius arquata*	Eurasian Curlew	Courlis cendré	Zarapito Real	Großer Brachvogel	Maçarico-real
I/II	*Numenius borealis*	Eskimo Curlew	Courlis esquimau	Zarapito Boreal, Zarapito Esquimal, Chorlo Polar	Eskimo-Brachvogel	Maçarico-esquimó

App	Taxon	English	Français	Español	Deutsch	Português
II	*Numenius phaeopus*	Whimbrel	Courlis corlieu	Zarapito Trinador, Chorlo Trinador	Regenbrachvogel	Maçarico-galego
I/II	*Numenius tenuirostris*	Slender-billed Curlew	Courlis à bec grêle	Zarapito de Pico Fino	Dünnschnabel-Brachvogel	Maçarico-de-bico-fino
II	*Nyctalus lasiopterus*	Greater Noctule Bat	Grand Noctule	Nóctulo Gigante	Riesenabend-segler	Morcego-arborícolagigante
II	*Nyctalus leisleri*	Leisler's Bat	Noctule de Leisler	Nóctulo Pequeño	Kleiner Abendsegler	Morcego-arborícola-damadeira, Morcego-arborícolapequeno
II	*Nyctalus noctula*	Noctule Bat	Noctule	Nóctulo Común	Abendsegler	Morcego-arborícolagrande
I/II	*Orcaella brevirostris*	Irrawaddy Dolphin	Dauphin de l'Irrawaddy	Delfín del Río Irrawaddy	Irrawadi Delphin	Golfinho-de-irawaddy
II	*Orcaella heinsohni* Formerly included in *Orcaella brevirostris*	Australian Snubfin Dolphin	Dauphin à aileron retroussé d'Australie	Delfín beluga de Heinsohn; delfín de aleta chata australiano	Australischer Stupsfinnendelfin	Golfomho-de-heinsohn
II	*Orcinus orca*	Killer whale,	Orca Epaulard, Orque	Orca	Schwertwal	Orca
I/II	*Oryx dammah*	Scimitar-horned Oryx	Oryx algazelle	Orix Cimitarra	Krummhornoryx, Säbelantilope	Órix-cimitarra
II	*Otaria flavescens*	South American Sea lion	Lion de mer d'Amérique du Sud	León marino sudamericano, lobo común, Léon marino austral	Südamerikanischer Seelöwe	Leão-marinho-sulamericano
I/II	*Otis tarda*	Great Bustard	Outarde barbue, Grande outarde	Avutarda	Großtrappe	Abetarda-comum
II	*Otomops madagascariensis* Formerly included in *Otomops martiensseni*					
II	*Otomops martiensseni*	Large-eared Free-tailed Bat			Martienssen Großohr-fledermaus	
I/II	*Oxyura leucocephala*	White-headed Duck	Erismature à tête blanche	Malvasía	Weißkopfruderente	Pato-de-rabo-alçado
II	*Pandion haliaetus*	Osprey, Fish Hawk	Balbuzard fluviatile, Balbuzard pêcheur	Aguila Pescadora	Fischadler	Águia-pescadora, Águia-pesqueira
I	*Pangasianodon gigas*	Giant Catfish	Silure de verre géant	Siluro Gigante	Riesenwels	Peixe-gato, Peixe-gatogigante-do-Mekong
I	*Pelecanoides garnotii*	Peruvian diving petrel	Puffinure de Garnot	Pato Yunco	Garnot-Lum-mensturmvogel	Petrel-mergulhadorperuano
I/II	Pelecanus crispus	Dalmatian Pelican	Pélican frisé, Pélican dalmate	Pelicano Ceñudo	Krauskopfpelikan	Pelicano-crespo
I/II	*Pelecanus onocrotalus*	White Pelican	Pélican blanc	Pelicano Vulgar	Rosapelikan	Pelicano-branco, Pelicano-comum
II	*Phalacrocorax nigrogularis*	Socotra Cormorant	Cormoran de Socotra	Cormorán de Socotra	Sokotrascharbe	Corvo-marinho-arábico
II	*Phalacrocorax pygmeus* Formerly listed as *Phalacrocorax pygmaeus*	Pygmy Cormorant	Cormoran pygmée	Cormorán Pigmeo	Zwergscharbe	Corvo-marinho-pigmeu
II	*Phalaropus fulicaria*	Grey Phalarope	Phalarope à bec large	Falaropo Picogrueso	Thorshühnchen	Falaropo-de-bico-grosso

App	Taxon	English	Français	Español	Deutsch	Português
II	*Phalaropus lobatus*	Red-necked Phalarope	Phalarope à bec étroit	Falaropo Picofino	Odinshühnchen	Falaropo-de-bico-fino
II	*Philomachus pugnax*	Ruff	Chevalier combattant, Combattant varié	Combatiente	Kampfläufer	Combatente
II	Phoca vitulina	Common Seal, Harbour Seal	Phoque commun, Phoque veau-marin	Foca Común	Seehund, Seerobbe	Foca-comum
II	*Phocoena dioptrica*	Spectacled Porpoise	Marsouin à lunettes	Marsopa de Anteojos	Brillenschweinswal	Phocoena dioptrica
II	*Phocoena phocoena*	Common Porpoise, Harbour Porpoise	Marsouin commun	Marsopa Común	Schweinswal, Braunfisch	Golfinho-de-óculos
II	*Phocoena spinipinnis*	Burmeister Porpoise	Marsouin de Burmeister	Marsopa Espinosa	Burmeister--Schweinswal	Boto-de-burmeister
II	*Phocoenoides dalli*	Dall's Porpoise	Marsouin de Dall	Marsopa de Dall	Dall-Hafensch-weinswal	Boto-de-dall, Toninhade-dall
II	*Phoebetria fusca*	Sooty Albatross	Albatros brun	Albatros Ahumado, Albatros Oscuro	Rußalbatros, Dunkelalbatros	Piau-preto
II	*Phoebetria palpebrata*	Light-mantled Sooty Albatross	Albatros fuligineux	Albatros Tiznado	Graumantel-Rußalbatros	Piau-de-costa-clara
I	*Phoenicopterus andinus* Formerly listed as *Phoenicoparrus andinus*	Andean Flamingo	Flamant des Andes	Parina Grande	Gelbfuß-flamingo	Flamingo-andino, Flamingo-dos-andes
I	*Phoenicopterus jamesi* Formerly listed as *Phoenicoparrus jamesi*	Puna Flamingo	Flamant des James	Parina Chica	Kurzschnabel-flamingo	Flamingo-de-james
II	*Phoenicopterus minor*	Lesser Flamingo	Petit flamant, Flamant nain	Flamenco Enano	Zwergflamingo	Flamingo-pequeno
II	*Phoenicopterus ruber*	Greater Flamingo	Flamant rose	Flamenco Común	Rosaflamingo	Flamingo, Flamingo-grande, Flamingo-rosa
I/II	*Physeter macrocephalus*	Sperm Whale	Cachalot	Ballena Esperma	Pottwal	Cachalote
II	*Pipistrellus kuhli*	Kuhl's Pipistrelle Bat	Pipistrelle de Kuhl	Murciélago de Borde Claro	Weißrandfled-ermaus	Morcego-de-kuhl
II	*Pipistrellus nathusii*	Nathusius's Pipistrelle Bat	Pipistrelle de Nathusius	Falso Murciélago Común	Rauhhautfld-ermaus	Morcego-de-Nathusius
II	*Pipistrellus pipistrellus*	Common Pipistrelle	Pipistrelle commune	Murciélago Común	Zwergfledermaus	Morcego-anão
II	*Pipistrellus savii*	Savi's Pipistrelle Bat	Pipistrelle de Savi	Murciélago Montañero	Alpenfledermaus	Morcego-de-savi
II	*Platalea alba*	African Spoonbill	Spatule d'Afrique	Espátula Africana	Rosenfußlöffler	Colhereiro-africano
II	*Platalea leucorodia*	Eurasian Spoonbill	Spatule blanche eurasienne	Espátula Blanca Löffler,	Euroasialöffler	Colhereiro-europeu
I	Platalea minor	Black-faced Spoonbill	Petite spatule	Espátula menor	Schwarzgesicht löffler	Colhereiro-comum, Colhereiro-europeu
I/II	*Platanista gangetica* gangetica Formerly listed as *Platanista gangetica*	Ganges River Dolphin, Blind River Dolphin	Platanista du Gange	Delfín del Río Ganges	Ganges-Delphin	Golfinho-do-ganges
II	*Plecotus auritus*	Brown Long-eared Bat	Oreillard brun	Orejudo Septentrional	Braunes Langohr	Morcego-orelhudocastanho
II	*Plecotus austriacus*	Grey Long-eared Bat	Oreillard gris	Orejudo Meridional	Graues Langohr	Morcego-orelhudocinzento
II	*Plectropterus gambensis*	Spur-winged Goose	Canard armé, Oie-armée de Gambie	Ganso Espolonado	Sporngans	Pato-ferrão
II	*Plegadis falcinellus*	Glossy Ibis	Ibis falcinelle	Morito	Sichler	Íbis-preta, Maçaricopreto

App	Taxon	English	Français	Español	Deutsch	Português
II	Pluvialis apricaria	Eurasian Golden Plover	Pluvier doré	Chorlito Dorado Europeo	Goldregenpfeifer	Tarambola-dourada
II	*Pluvialis squatarola*	Grey Plover	Pluvier argenté	Chorlito Gris, Chorlo Artico	Kiebitzregenpfeifer	Tarambola-cinzenta, Batuiruçu-de-axila-preta
II	*Podiceps auritus*	Slavonian Grebe, Horned Grebe	Grèbe esclavon	Zampullín Cuellirroio	Ohrentaucher	Mergulhão-de-pescoçocastanho
II	*Podiceps grisegena grisegena*	Red-necked Grebe	Grèbe jougris	Somormuio Cuellirroio	Rothalstaucher	Mergulhão-de-pescoçoruivo
I/II	*Podocnemis expansa*	Arrau Turtle, South American River Turtle	Tortue de l'Amazone	Tortuga Arrau, Tortuga Fluvial, Charapa	Arrauschildkröte	Tartaruga-de-amazônia
I/II	*Polysticta stelleri*	Steller's Eider	Eider de Steller	Eider Menor	Scheckente	Eider-de-steller
II	*Polystictus pectoralis pectoralis*	Bearded Tachuri	Tyranneau barbu	Tachuri, Tachuri canela	Schmalschwanz-tyrann	Tricolino-canela
I/II	*Pontoporia blainvillei*	La Plata Dolphin, Franciscana	Dauphin de la Plata	Delfín del Plata, Franciscana	La-Plata-Delphin	Franciscana, Golfinhodo-rio-da-plata
II	*Porzana parva parva*	Little Crake	Marouette poussin	Polluela Bastarda Kleines	Sumpfhuhn	Franga-d'agua-bastarda
II	*Porzana porzana*	Spotted Crake	Marouette ponctuée	Polluela Pintoja	Tüpfelsumpfhuhn	Franga-d'agua-grande
II	*Porzana pusilla intermedia*	Baillon's Crake	Marouette de Baillon	Polluela Chica	Zwergsumpfhuhn	Franga-d'agua-pequena
II	*Procapra gutturosa*	Mongolian or Whitetailed gazelle	Gazelle à queue blanche, Gazelle de Mongolie	Gacela de Mongolia, zeren	Mongolische Gazelle, Mongol-eigazelle	Gazela da Mongólia
II	*Procellaria aequinoctialis* Includes Procellaria *aequinoctialis conspicillata* previously listed separately as *Procellaria conspicillata*	White-chinned Petrel Spectacled Petrel	Puffin à menton blanc Puffin à lunettes	Pardela Gorgiblanca, Petrel de Barba Blanca Petrel de Antifaz	Weißkinn-Sturmvogel Weißkinn-Sturmvogel	Pardela-preta Pardela-de-óculos
II	*Procellaria cinerea*	Grey Petrel	Puffin gris	Pardela Gris	Grausturmvogel	Pardela-cinza
II	*Procellaria parkinsoni*	Black Petrel	Puffin de Parkinson	Pardela de Parkinson	Schwarzsturmvogel	Pardela-de-Parkinson
II	*Procellaria westlandica*	Westland Petrel	Puffin de Westland	Pardela de Westland	Westlandsturm-vogel	
II	*Psephurus gladius*	Chinese Paddlefish, Chinese Swordfish, White Sturgeon	Spatule du Chang jiang	Pez espátula chino	Schwertstör	Peixe-espada-doyangzi, peixe-espátula
II	*Pseudocolopteryx dinellianus*	Dinelli's Doradito	Doradite de Dinelli	Doradito pardo	Dinellisumpf-tyrann	Tricolino-pardo
II	*Pseudoscaphirhynchus fedtschenkoi*	Syr-Dar Shovelnose	Nez-pelle du Syr Daria		Syr-Darja-Schaufelstör	
II	*Pseudoscaphirhynchus hermanni*	Small Amu-Dar Shovelnose	Petit nez-pelle de l'Amou daria	Esturión enano	Kleiner Amu-Darja-Schaufelstör	
II	*Pseudoscaphirhynchus kaufmanni*	Large Amu-Dar Shovelnose, False Shovelnose, Shovelfish	Grand nez-pelle de l'Amou Daria		Großer Pseudo-schaufelstör	
I	*Pterodroma atrata*	Henderson Petrel	Pétrel de Henderson	Fardela de Henderson	Henderson Sturmvogel	Petrel-de-henderson
I	*Pterodroma cahow*	Cahow, Bermuda Petrel	Pétrel cahow	Petrel Cahow	Bermuda-sturmvogel	Freira-das-bermudas
I	*Pterodroma phaeopygia*	Dark-rumped Petrel, Hawaiian Petrel, Galapagos Petrel	Pétrel à croupion sombre, Pétrel des Hawaii	Petrel Hawaiano	Hawaiisturmvogel	Petrel-do-havaí

App	Taxon	English	Français	Español	Deutsch	Português
I	*Pterodroma sandwichensis* Formerly included in *Pterodroma phaeopygia* (s.l.)	Dark-rumped Petrel, Hawaiian Petrel, Uau	Pétrel à croupion sombre, Pétrel des Hawai	Petrel Hawaiano	Hawaiisturmvogel	Petrel-do-havaí
I	*Puffinus creatopus*	Pink-footed Shearwater	Puffin à pieds roses	Fardela blanca, Fardela de vientre blanco	Blassfuß Sturmvogel	Pardela-de-patasrosadas
I	*Puffinus mauretanicus*	Balearic shearwater	Puffin des Baléares	Pardela balear	Balearan sturmvogel	Pardela-domediterrâneo
II	*Recurvirostra avosetta*	Pied Avocet	Avocette élégante	Avoceta Común	Säbelschnäbler	Alfaiate
II	*Rhincodon typus*	Whale Shark	Requin-baleine	Tiburón Ballena, Pez Dama	Walhai, Rauhhai	Tubarão-baleia
II	*Rhinolophus blasii*	Blasius' Horseshoe Bat	Rhinolophe de Blasius	Murciélago Dalmata de Herradura	Blasius-Hufeisennase	Morcego-ferradura-denariz-de-sela
II	*Rhinolophus euryale*	Mediterranean Horseshoe Bat	Rhinolophe euryale	Murciélago Mediterráneo de Herradura	Mittelmeer-Hufeisennase	Morcego-de-ferradura mediterrânico
II	*Rhinolophus ferrumequinum*	Greater Horseshoe Bat	Grand rhinolophe fer à cheval	Murciélago Grande de Herradura	Große Hufeisennase	Morcego-de-ferraduragrande
II	*Rhinolophus hipposideros*	Lesser Horseshoe Bat	Petit rhinolophe fer à cheval	Murciélago Pequeño de Herradura	Sanborns Fledermaus, Kleine Hufeisennase	Morcego-de-ferradura-pequeno
II	*Rhinolophus mehelyi*	Mehely's Horse-shoe Bat	Rhinolophe de Mehely	Murciélago de Herradura de Mehely	Mehely-Hufeisennase	Morcego-de-ferradur-amourisco
II	*Rynchops flavirostris*	African skimmer	Bec-en-ciseaux d'Afrique	Picotijera africano	Braunmantel-Scherenschnabel Afrikanischer Scherenschnabel	Bico-de-tesoura-africano
II	*Saiga tatarica*	Saiga antelope	Saïga	Saiga	Saiga	Saiga
II	*Sarkidiornis melanotos*	Comb Duck	Canard casqué, Canard à bosse bronzé	Pato Crestudo	Glanzente	Pato-de-crista
I/II	*Sarothrura ayresi*	Whitewinged Flufftail	Râle à miroir	Polluela Especulada	Spiegelralle	Frango-d'agua-d'asabranca
II	*Sarothrura boehmi*	Streaky-breasted Flufftail	Râle de Boehm	Polluela de Boehm	Böhmralle	Frango-d'agua-deboehm
I	*Serinus syriacus*	Syrian Serin	Serin syriaque	Serín de Siria	Zederngirlitz	Milheirinha-do-levante
II	*Somateria mollissima*	Common Eider	Eider à duvet	Eider Común	Eiderente	Eider-edredão
II	*Somateria spectabilis*	King Eider	Eider à tête grise	Eider Real	Prachteiderente	Eider-real
II	*Sotalia fluviatilis*	Tucuxi, Bouto Dolphin	Sotalia, Dauphin de l'Amazon	Delfín del Amazonas	Amazonas-Sotalia	Boto-cinza, Tucuxi
II	*Sotalia guianensis* Formerly included in *Sotalia fluviatilis*	Costero				Boto-cinza
II	*Sousa chinensis*	Indo-Pacific Hump-backed Dolphin Chinese White Dolphin	Dauphin blanc de Chine	Delfín Blanco de China	Chinesischer Weißer Delphin	Golfinho-corcunda-doindopacífico
II	*Sousa teuszii*	Atlantic Hump-backed Dolphin, Cameroon Dolphin	Dauphin du Cameroun	Delfín Jorobado del Atlántico	Kamerunfluss-Delphin	Golfinho-corcunda-doatlântico
II	*Spheniscus demersus*	African Penguin	Manchot du Cap	Pingüino del Cabo	Brillenpinguin	Pinguim-africano, Pinguim-do-cabo
I	*Spheniscus humboldti*	Humboldt Penguin	Manchot de Humboldt	Pingüino de Humboldt	Humboldtpinguin	Pinguim-de-humboldt

App	Taxon	English	Français	Español	Deutsch	Português
I/II	*Sporophila cinnamomea*	Chestnut Seedeater	Sporophile cannelle	Capuchino Corona Gris		Caboclinho-de-chapéucinzento
I/II	*Sporophila hypochroma*	Rufous-rumped Seedeater	Sporophile à croupion roux	Capuchino Castaño	Rotbürzelp-fäffchen	Caboclinho-de-sobreferrugem
I/II	*Sporophila palustris*	Marsh Seedeater	Sporophile des marais	Capuchino pecho blanco	Sumpfpfäffchen	Caboclinho-de-papobranco
II	*Sporophila ruficollis*	Dark-throated Seedeater	Sporophile à gorge sombre paraguai,	Capuchino garganta café	Schwarz-kehlpfäffchen	Caboclinho-Caboclinho-do-papoescuro, Cabo-clinho-dopapopreto
I/II	*Sporophila zelichi*	Zelich's Seedeater	Sporophile de Naroski	Capuchino de Collar	Zelichpfäffchen	Caboclinho-de-coleirabranca
II	*Squalus acanthias*	Spiny dogfish	Aiguillat commun	mielga, galludos, cazón espinozo, tiburón espinozo, espineto, espinillo, tollo, tollo de cachos	Dornhai	Cação-Bagre, Cação-deespinho, Galhudomalhado
II	*Stenella attenuata*	Pantropical Spotted Dolphin, Bridled Dolphin	Dauphin tacheté de Pantropical	Estenela Moteada	Schlankdelphin	Golfinho-pintadotropical, Golfinhopintado-pantropical
II	*Stenella clymene*	Clymene dolphin	Dauphin Clymène	Delfín clymene	Clymene Delphin	Golfinho-clímene, Golfinho-fiandeiro-debico-curto
II	*Stenella coeruleoalba*	Striped Dolphin, Bluewhite Dolphin	Dauphin bleu et blanc, Dauphin rayé	Delfín Listado	Blauweißer Delphin	Golfinho-riscado, Toninha-riscada
II	*Stenella longirostris*	Spinner Dolphin	Dauphin à ventre rose	Delfín Tornillon, Delphín Churumbelo	Ostpazifischer Delphin	Golfinho-rotador, Golfinho-fiandeiro-debico-comprido
II	*Sterna albifrons*	Little Tern	Sterne naine	Charancito	Zwergseeschwalbe	Andorinha-do-mar-anã, trinta-réis-miúdo
II	*Sterna balaenarum*	Damara Tern	Sterne des baleiniers	Charrancito de Damara	Damaraseeschwalbe	Gaivina-de-damara
II	*Sterna bengalensis*	Lesser Crested Tern	Sterne voyageuse	Charrán Bengalés	Rüppelseeschwalbe	Gaivina-de-bico-laranja, Garajau-bengalense
II	*Sterna bergii*	Great Crested Tern	Sterne huppée	Charrán de Berg	Eilseeschwalbe	Gaivina-de-bicoamarelo, Garajau-debico-amarelo
I	*Sterna bernsteini*	Chinese Crested Tern	Sterne d'Orient	Pegaza de oriente	Bernsteinsee-schwalbe	
II	*Sterna caspia*	Caspian Tern	Sterne caspienne	Pagaza Piquirroja	Raubseeschwalbe	Gaivina-de-bicovermelho
II	*Sterna dougallii*	Roseate Tern	Sterne de Dougall	Charrán Rosado	Rosenseeschwalbe	Andorinha-do-marrósea, Gaivina-rosada, Garajau-rosado
II	*Sterna hirundo hirundo*	Common Tern	Sterne pierregarin	Charrán Común, Gaviotín Golondrina	Flussseeschwalbe	Andorinha-do-marcomum, Garajaucomum
I	*Sterna lorata*	Peruvian tern	Sterne du Pérou	Gaviotín chico	Peruseeschwalbe	Garajau-real

App	Taxon	English	Français	Español	Deutsch	Português
II	*Sterna maxima albidorsalis*	Royal Tern	Sterne royale	Charrán Real	Königsseeschwalbe	
II	*Sterna nilotica nilotica*	Gull-billed Tern	Sterne hansel	Pagaza Piconegra	Lachseeschwalbe	Gaivina-de-bico-preto
II	*Sterna paradisaea*	Arctic Tern	Sterne arctique	Charrán Artico, Gaviotín Artico	Küstenseeschwalbe	Andorinha-do-marártica, Trinta-réis-ártico
II	*Sterna repressa*	White-cheeked Tern	Sterne à joues blanches	Charrán Cariblanco	Weißwangen seeschwalbe	Gaivina-cinzenta, Gaivina-arábica
II	*Sterna sandvicensis sandvicensis*	Sandwich Tern	Sterne caugek	Charrán Patinegro	Brandseeschwalbe	Trinta-réis-de-bando
II	*Sterna saundersi*	Saunder's Tern	Sterne de Saunders	Charrán de Saunders	Orientseeschwalbe	Chilreta-de-Saunders
II	*Streptopelia turtur turtur*	Turtle dove	Tourterelle des bois, Tourterelle européenne	Tórtola europea	Turteltaube	Rola-comum
I	*Synthliboramphus wumizusume*	Japanese Murrelet, Crested Murrelet	Guillemot du Japon	Mérgulo Japonés	Japanalk	
I	*Tadarida brasiliensis*	Mexican Free-tailed Bat	Chauve-souris à queue-libre du Mexique	Rabudo Mejicano	Brasilianische Bulldogfledermaus, Guano Fledermaus	Morcego-comum, Morcego-das-casas
II	*Tadarida insignis* Formerly included in *Tadarida teniotis*					
II	*Tadarida latouchei* Formerly included in *Tadarida teniotis*	La Touche's Free-tailed Bat				
II	*Tadarida teniotis*	European Free-tailed Bat	Molosse de Cestoni	Murciélago Rabudo	Europäische Bulldogfledermaus	Morcego-rabudo
II	*Tadorna cana*	South African Shelduck	Tadorne à tête grise	Tarro Sudafricano	Graukopfkasarka	Tadorna-africana
II	*Tadorna ferruginea*	Ruddy Shelduck	Tadorne casarca	Tarro Canelo	Rostgans	Pato ferrugíneo
II	*Tadorna tadorna*	Common Shelduck	Tadorne de Belon	Tarro Blanco	Brandgans	Pato-branco
II	*Thalassornis leuconotus*	White-backed Duck	Canard à dos blanc, Dendrocygne à dos blanc	Pato Dorsiblanco	Weißrücken-Pfeifgans, Weißrückenente	Pato-de-dorso-branco
II	*Threskiornis aethiopicus aethiopicus*	Sacred Ibis	Ibis sacré	Ibis Sagrado	Heiliger Ibis	Íbis-sagrada
II	*Trichechus inunguis*	Amazonian Manatee	Lamantin d'Amazonie ou du Brésil	Manatí amazónico, Vaca marina	Amazonas-Manati; Flußmanati	Peixe-boi-da-amazônia, Peixe-boi-amazônico
I/II	*Trichechus manatus*	Manatee	Lamantin	Manatí	Seekuh	Peixe-boi-marinho
II	*Trichechus senegalensis*	West African Manatee	Lamantin ouest-africain	Manatí de África Occidental	Afrikanischer Manati	Peixe-boi-africano
II	*Tringa cinerea*	Terek Sandpiper	Bargette de Terek, Chevalier bargette	Andarríos del Terek	Terekwasserläufer	Maçarico-sovela
II	*Tringa erythropus*	Spotted Redshank, Dusky Redshank	Chevalier arlequin	Archibebe Oscuro	Dunkler Wasserläufer	Perna-vermelha-escuro
II	*Tringa glareola*	Wood Sandpiper	Chevalier sylvain	Andarríos Bastardo	Bruchwasserläufer	Maçarico-bastardo, Maçarico-de dorsomalhado
I/II	*Tringa guttifer*	Spotted Greenshank, Nordmann's Greenshank	Chevalier tacheté		Tüpfel-Grünschenkel	Perna-verde-pintado
II	*Tringa hypoleucos*	Common Sandpiper	Chevalier guignette	Andarríos Chico	Flussuferläufer	Maçarico-das-rochas
II	*Tringa nebularia*	Common Greenshank	Chevalier aboyeur	Archibebe Claro	Grünschenkel	Perna-verde-comum

App	Taxon	English	Français	Español	Deutsch	Português
II	*Tringa ochropus*	Green Sandpiper	Chevalier cul-blanc	Andarríos Grande	Waldwasserläufer	Maçarico-bique-bique, Pássaro-bique-bique
II	*Tringa stagnatilis*	Marsh Sandpiper	Chevalier stagnatile	Archibebe Fino	Teichwasserläufer	Perna-verde-fino
II	*Tringa totanus*	Common Redshank	Chevalier gambette	Archibebe Común	Rotschenkel	Perna-vermelha-comum, Maçarico-de-pernavermelha, Cacongo
I/II	*Tryngites subruficollis*	Buff-breasted Sandpiper	Bécasseau rousset	Playerito Canela	Grasläufer	Maçarico-acanelado, Pilrito-canela
II	*Tursiops aduncus*	Indian or Bottlenose Dolphin	Grand dauphin de l'océan Indien	Delfín Mular del Indico	Großer Tümmler des Indischen Ozeans	Golfinho-nariz-degarrafa-do-índico, Golfinho-flíper-comum
II	*Tursiops truncatus*	Bottlenosed Dolphin	Grand dauphin, Souffleur, Tursion	Delfín Mular	Großer Tümmler	Golfinho-nariz-degarrafa, Golfinho-roaz, Golfinho-roazcorvineiro
I	*Tursiops truncatus ponticus*	Bottle-nosed dolphin	Grand dauphin	Delfín Mular	Schwarzmeer-	Golfinho-nariz-degarrafa-do-mar-negro
I	*Uncia uncia* Formerly listed as *Panthera uncia*	Snow Leopard	Panthère des neiges, Irbis, Once	Pantera de las Nieves, Leopardo Blanco	Schneeleopard, Irbis	Leopardo-das-neves
II	*Vanellus albiceps*	White-headed Lapwing	Vanneau à tête blanche	Avefría Coroniblanca	Langspornkiebitz, Weißscheitel-kiebitz	Abibe-de-coroa-branca, Barbilhão-de-golabranca
II	*Vanellus coronatus*	Crowned Lapwing	Vanneau couronné	Avefría Coronada	Kronenkiebitz	Abibe-coroado,
I/II	*Vanellus gregarius* Formerly listed as *Chettusia gregaria*	Sociable Plover	Vanneau sociable	Avefria Sociable	Steppenkiebitz	Abibe-sociável
II	*Vanellus leucurus*	White-tailed Plover	Vanneau à queue blanche	Avefría Coliblanca	Weißschwanz-steppenkiebitz	Abibe-de-cauda-branca
II	*Vanellus lugubris*	Wattled Lapwing	Vanneau demi-deuil, Vanneau terne	Avefría Lúgubre	Trauerkiebitz	Abibe-d'asa-negrapequeno
II	*Vanellus melanopterus*	Black-winged Lapwing	Vanneau à ailes noires	Avefría Lugubroide	Schwarzflügel-kiebitz	Abibe-d'asa-negra
II	*Vanellus senegallus*	Senegal Lapwing	Vanneau du Sénégal	Avefría Senegalesa	Senegalkiebitz	Abibe-carunculado, Abibe-do-senegal, Barbilhão-amarelo
II	*Vanellus spinosus*	Spur-winged Plover	Vanneau à éperons	Avefría Espinosa	Spornkiebitz	Abibe-esporado
II	*Vanellus superciliosus*	Brown-chested Lapwing	Vanneau caronculé, Vanneau à poitrine châtaine	Avefría Pechirrufa	Rotbrustkiebitz	
II	*Vanellus vanellus*	Northern Lapwing	Vanneau huppé	Avefría Europea	Kiebitz	Abibe-comum
II	*Vespertilio murinus*	Parti-coloured Bat	Sérotine bicolore	Murciélago Bicolor	Zweifarbfledermaus	Morcego-bicolor
I/II	*Vicugna vicugna*	Vicugna	Vigogne	Vicuña	Vikunja	Vicunha
I/II	*Zoothera guttata*	Spotted ground thrush	Grive tachetée	Zorzal Moteado	Nataldrossel	Tordo-malhado

The following Families and Genera are listed in Appendix II without naming each species individually. Some of the species belonging to the higher taxon are also listed separately on Appendix I, and where this is the case, first column of this table is annotated I/II (this does not mean that all species belonging to the higher taxon are listed in both Appendices)

App	Taxon	English	Français	Español	Deutsch	Português
II	*Rhinolophidae*	Horseshoe bats	Rhinolophidés	Murciélagos grandes de herradura	Hufeisennasen	Morcego-de-ferradura
II	*Vespertilionidae*	Evening bats	Vespertilionidés	Vespertiliónidos	Glattnasen	Vespertilionídeos
I/II	*Phoenicopteridae*	Flamingos	Phoenicoptéridés	Flamencos	Flamingos	Flamingos
I/II	*Anatidae*	Ducks	Anatidés	Anátidas	Entenvögel	Anatídeos
II	*Cathartidae*	New World vultures	Vautours du Nouveau Monde	Buitres del Nuevo Mundo	Neuweltgeier	Abutres-do-novomundo
I/II	*Accipitridae*	Hawks, eagles, kites, harriers and Old World vultures	Milans, aigles, busards, vautours	Águilas, milanos y buitres del Viejo Mundo	Habichte, Adler, Altweltgeier	Accipitrídeos
I/II	*Falconidae*	Falcons and caracaras	Faucons, fauconnets, crécerelles	halcónes y caracaras o caranchos	Falken, Geierfalken	Falconídeos
I/II	*Grus spp*	Cranes	Grues	Grullas	Kraniche	Grous
II	*Recurvirostridae*	Avocets and stilts	Avocettes et échasses	Avocetas	Säbelschnäbler	Alfaiates,Perna-longas
I/II	*Charadriidae*	Plovers, dotterels, and lapwings	Les vanneaux, les pluviers, les échasses et les gravelots.	Chorlos, chorlitos, avefrías y chorlitejos	Regenpfeifer	Abibes, Tarambolas, Borrelhos
I/II	*Scolopacidae* Includes the sub-family *Phalaropodinae* Formerly listed as the family *Phalaropodidae*	Sandpipers, curlews and godwits	Bécasses, bécassines, bécassins (ou limnodromes), barges, courlis, maubèche (ou bartramie), chevaliers, tournepierres, bécasseaux, combattant et phalaropes.	Escolopácidos	Schnepfenvögel	Maçaricos, Narcejas, Fuselos
I/II	*Muscicapidae* Includes the sub-family *Sylviinae*, Formerly listed as the family *Sylviidae*	Old World flycatchers	Muscicapidés	Muscicapidae	Fliegenschnäpper	Muscicapídeos
I/II	*Cheloniidae*	Sea turtles	Tortues marines	Tortugas marinas	Meeresschildkröten	Tartarugas marinhas
I/II	*Dermochelyidae*	Leatherback sea turtle	Tortue luth	Tortuga laúd	Lederschildkröte	Tartarugas de couro, Tartaruga-gigante

Appendix II: Role of IUCN

IUCN has been connected to CMS throughout the Convention's existence, not just by their shared history but also by a like-minded philosophy and approach to global environmental governance.

In the 1960s, before the pivotal Stockholm Conference which took place in 1972, IUCN was issuing alarm calls about the status of migratory species. When the UN Conference called for an international instrument to protect endangered migratory species, the IUCN's eleventh General Assembly endorsed the concept as a priority. It was thanks to the dedication of a small number of enthusiasts to ensure that the idea became fact – notably Wolfgang Burhenne, Françoise Burhenne-Guilmin and the late Cyrille de Klemm – the masterminds along with the German Government behind the first draft text that formed the basis of the Convention finally adopted in 1979. They developed an innovative and flexible text, which made CMS a "framework" convention under which species-specific, regional Agreements could be concluded. The German Government steered this text through a series of negotiations culminating in a diplomatic conference in Bonn in 1979 where CMS was concluded.

Having served as "midwife" to the Convention itself, IUCN has fulfilled a similar role for many of the Convention's daughter Agreements and Memoranda of Understanding (MoU). These have all benefited from the direct involvement of IUCN's regional offices and the Specialist Groups of the Species Survival Commission. The IUCN's Marine Programme provides an excellent example of how IUCN's concerns overlap with those of CMS, as many CMS species live by or in the sea. The Environmental Law Centre (ELC) helped draw up the African-Eurasian Waterbird Agreement (AEWWB) and is represented on AEWA's Technical Committee. IUCN's ELC also negotiated the Wadden Sea Seal Agreement and provided technical advice in drafting both ASCOBANS and ACCOBAMS.*

More recently IUCN has assisted with the development of ACAP (the Albatross and Petrel Agreement) and the MoU for marine turtles on the Atlantic Coast of Africa and Indian Ocean and South-East Asia. IUCN's Species Specialist Groups often provide valuable technical input into CMS initiatives, by helping to prepare Action Plans for a number of priority species, including the African Elephant. The IUCN's African Elephant Specialist Group (AfESG) has been particularly closely involved in the CMS efforts to secure the conclusion of a Memorandum of Understanding between the range states. Regular discussions take place between the Executive Secretary and the Chair of the IUCN Species Survival Commission.

Key facts

- Founded in 1948 as the International Union for the Protection of Nature (or IUPN), the organisation changed its name in 1956 to the International Union for Conservation of Nature and Natural Resources (IUCN). The name IUCN was adopted in 2008 (UICN in French).
- The Union has over 1,000 staff working in 62 different countries. Its headquarters are at Rue Mauverney 28, Gland, 1196 Switzerland.
- The Union's unique network consists of 82 state members, 11 government agencies and 800 non-governmental organisations. It can call upon the expertise and commitment of over 10,000 scientists from 181 countries.

** see page 62*

Appendix III: Global Biodiversity Instruments

Six global conventions and treaties focus on biodiversity issues: the Convention on Biological Diversity (CBD), the Convention on International Trade in Endangered Species of Wild Fauna and Flora (CITES), the Convention on Conservation of Migratory Species of Wild Animals (CMS), the International Treaty on Plant Genetic Resources for Food and Agriculture (ITPGRFA), the Ramsar Convention on Wetlands (Ramsar), and the World Heritage Convention (WHC). Each one works to reach common goals in conservation and sustainable use. They have evolved a range of methods (site, species and/or ecosystem-based) and operations such as work programmes, certification and licensing, regional agreements, site listings and financial instruments. While each is independent and answerable to its own Parties, the issues addressed overlap, and these links provide a basis for cooperation for both monitoring and implementation.

CBD 1992 (413 Saint Jacques Street, Suite 800, Montreal, Quebec, Canada H2Y 1N9)
The objectives of CBD are the conservation of biological diversity, the sustainable use of its components, and the fair and equitable sharing of the benefits arising from commercial and other utilization of genetic resources.

CITES 1975 (International Environment House, Chemin des Anémones, CH-1219 Châtelaine, Geneva, Switzerland)
CITES aims to ensure that international trade in wild animals and plants does not threaten the species' survival. Through its three appendices, the Convention accords protection to more than 30,000 plant and animal species.

CMS 1979 (UN Campus Bonn, Hermann-Ehlers-Str. 10, 53113 Bonn, Germany)
CMS aims to conserve terrestrial, aquatic and avian migratory species throughout their range. Parties to CMS provide strict protection for the most endangered migratory species, by concluding regional multilateral agreements for the conservation and management of specific species or categories of species, and by undertaking cooperative research and conservation activities.

ITPGRFA 2001 (Viale delle Terme di Caracalla, 1 00153 Rome, Italy)
The Treaty was approved by the Food and Agriculture Conference in November 2001. The Treaty is vital in ensuring the continued availability of the plant genetic resources that countries will need to feed their people. We must conserve for future generations the genetic diversity that is essential for food and agriculture.

Ramsar 1971 (Rue Mauverney 28, CH-1196 Gland, Switzerland)
The Ramsar Convention provides the framework for national action and international co-operation for the conservation and wise use of wetlands and their resources. The Convention covers all aspects of wetland conservation and wise use.

WHC 1972 (7, place de Fontenoy, 75352 Paris 07 SP, France)
The primary mission of WHC is to identify and conserve the world's cultural and natural heritage, by drawing up a list of sites whose value should be preserved for all humanity and to ensure their protection through co-operation.

Key facts

- CBD has 192 Parties, World Heritage Convention (WHC) 186, CITES 173, Ramsar 158 and ITPGRFA 119.
- CMS joined the "Century Club" at the end of 2006. Membership has since reached 113.
- CBD, CITES and CMS are administered by UNEP.
  See: http://www.biodiv.org/cooperation/related-conventions/blg.shtml.

Appendix IV: CMS Partnerships

AMMPA www.ammpa.org
The Alliance of Marine Mammal Parks and Aquariums is an international association representing marine life parks, aquariums, zoos, research facilities and professional organisations dedicated to the highest standards of care for marine mammals and to their conservation in the wild through public education, scientific study and wildlife presentations.

BLI www.birdlife.org
BirdLife International is a global Partnership of conservation organisations that strives to conserve birds, their habitats and global biodiversity, working with people towards sustainability in the use of natural resources. BirdLife Partners operate in over one hundred countries and territories worldwide and collaborate on regional work programmes in every continent.

CIC www.cic-wildlife.org
The International Council for Game and Wildlife Conservation is a politically independent advisory body internationally active on a non-profit basis. With its scientific capacity, the CIC assists governments and environmental organisations in maintaining natural resources by sustainable use. The acronym CIC comes from the organisation's French name Conseil International de la Chasse et de la Conservation du Gibier.

FZS www.zgf.de
The Frankfurt Zoological Society of 1858 (ZGF) is an internationally active conservation organisation based in Frankfurt, Germany. The ZGF aims to conserve biodiversity globally, especially in East Africa, going back to the activities of Professor Dr. Bernhard Grzimek, a famous German naturalist. Currently the ZGF supports about 70 conservation projects in 30 countries, as well as the Zoological Garden Frankfurt.

GNF www.globalnature.org
The Global Nature Fund (GNF) is an NGO which was founded in early 1998 with the objective to foster the protection of water, nature and the environment as well as wildlife and biodiversity. GNF's work consists mainly of: initiating and carrying out nature/environment protection projects to preserve the fauna, protect migratory species, their habitats and their migration routes; the implementation of model projects for the promotion of sustainable economy; publications and organisation of events dealing with the protection of nature and the environment; and supporting international conventions on species conservation.

GBIF www.gbif.org
The Global Biodiversity Information Facility was founded in 2001 and enables free and open access to biodiversity data online. GBIF is an international government-initiated and funded initiative focused on making biodiversity data available to all and anyone, for scientific research, conservation and sustainable development.

ICF www.savingcranes.org
The International Crane Foundation is an NGO, founded in 1973 by two Cornell University graduate students, George Archibald and Ron Sauey who while studying crane behaviour became aware of the intense pressures on the birds' remaining populations. ICF has been one of CMS's main partners in the MoU concerning conservation measures for the Siberian Crane since the MoU was concluded in 1993. ICF's headquarters are located in Baraboo, Wisconsin, in the USA.

IFAW www.ifaw.org
The International Fund for Animal Welfare (IFAW) is one of the largest animal welfare and conservation charities in the world. Its mission is "to improve the welfare of wild and domestic animals throughout the world by reducing commercial exploitation of animals, protecting wildlife habitats, and assisting animals in distress".

The work of IFAW's global team of campaigners, legal and political experts and scientists is concentrated in three areas: reducing commercial exploitation of wild animals; protecting wildlife habitats; and providing emergency relief to animals in distress.

IWC www.iwcoffice.org
The International Whaling Commission (IWC) was set up under the International Convention for the Regulation of Whaling which was signed in Washington DC on 2nd December 1946. The purpose of the Convention is to provide for the conservation of whale stocks and thus make possible the orderly development of the whaling industry.

NOAA www.noaa.gov
The National Oceanic and Atmospheric Administration (NOAA) is a US federal agency set up under the Department of Commerce focused on the condition of the oceans and the atmosphere. NOAA dates back to 1807, when the USA's first scientific agency, the Survey of the Coast, was established. Since then, NOAA has evolved to become one of the leading scientific and environmental authorities internationally.

SCF (Sahara Conservation Fund)
The Sahara Conservation Fund (SCF) is an international non-governmental organisation established in 2004 to conserve the wildlife of the Sahara and bordering Sahelian grasslands. SCF's vision is of a Sahara that is well conserved and managed, in which ecological processes function naturally, with plants and animals existing in healthy numbers across their historical range; a Sahara that benefits all its inhabitants and users and where support for its conservation comes from stakeholders across all sectors of society. To implement its mission, SCF forges partnerships between people, governments, the world zoo and scientific communities, international conventions, NGOs and donor agencies.

WAZA www.waza.org
The World Association of Zoos and Aquariums is the 'umbrella' organisation for the world zoo and aquarium community. Its members include leading zoos and aquariums, and regional and national Associations of Zoos and Aquariums, as well as some affiliate organisations, such as zoo veterinarians or zoo educators, from all around the world. WAZA represents its members in other international bodies such as IUCN or Conferences of the Parties (COPs) to global Conventions, such as CITES, CBD or CMS.

WCS www.wcs.org
The Wildlife Conservation Society is a US-based NGO that endeavours to save wildlife and wild lands though careful use of science, conservation action, education and the integration of urban wildlife parks. It runs a large international conservation programme and is based at the Bronx Zoo in New York.

WDCS www.wdcs.org
WDCS, the Whale and Dolphin Conservation Society, is a charity dedicated to the conservation and welfare of all whales, dolphins and porpoises. Established in 1987, WDCS is staffed by over 70 people, along with many volunteers, located in its offices in Argentina, Australia, Austria, Germany, the UK and the US. It was a partner in the Year of the Dolphin campaigns in 2007 and 2008.

Wetlands International www.wetlands.org
Wetlands International is a global NGO dedicated to the conservation and wise use of wetlands. The group acts globally, regionally and nationally to achieve the conservation and wise use of wetlands, to benefit biodiversity and human well-being. Its head office is in Wageningen, the Netherlands.

WHMSI
www.fws.gov/international/whmsi/whmsi_Eng.htm
The Western Hemisphere Migratory Species Initiative (WHMSI) seeks to contribute significantly to the conservation of the migratory species of the Western Hemisphere by strengthening communication and cooperation among nations, international conventions and civil society, and by expanding constituencies and political support. The initiative includes all migratory species, covering taxa as diverse as birds, marine turtles, marine and terrestrial mammals, fish and invertebrates.

ZSL www.zsl.org
The Zoological Society of London (ZSL) is a charity devoted to the worldwide conservation of animals and their habitats. ZSL scientists in the laboratory and the field, animal management teams at Regent's Park and Whipsnade zoos and veterinarians contribute wide-ranging skills and experience to both practical conservation and to the scientific research that underpins this work.

Appendix V: CMS Small Grants Programme

The Small Grants Programme (SGP) plays a significant role in promoting CMS initiatives for a number of taxa, mainly in developing countries. It has been the Convention's main tool in supporting Concerted Actions of Annex I species like the Sahelo-Saharan antelopes, marine turtles of the Indian Ocean and South East Asia and along the Atlantic Coast of Africa, small cetaceans in South East Asia and in tropical West Africa, the Siberian Crane, the Ruddy-headed Goose and Andean flamingos.

The SGP has proved successful at levering further funding from other donors and acting as 'seed corn' money leading to more ambitious conservation initiatives. The SGP has also helped raise CMS's profile and establish the Convention's reputation as a practical international conservation instrument in the countries and regions benefiting from the funding.

The Scientific Council is entrusted with the task of identifying suitable projects and establishing a priority list of actions. Individual members of the Council are then assigned to monitor each project. At its thirteenth meeting, 19 proposals were endorsed and recommended for funding by the Council. Four of the projects covered marine mammals, three terrestrial mammals, nine birds, two fish and one marine turtle. The geographic spread was four projects in Europe, four in Latin America and the Caribbean, four in Africa, one in Oceania and the Pacific and four in Asia, while two crossed continental boundaries.

Funding: Up until 2005 all the projects were financed through authorised withdrawals against the CMS Trust Fund's accumulated surplus. Since COP8 and for the foreseeable future however, the SGP will have to rely almost exclusively on voluntary contributions from the Parties – either donated generally or earmarked for specific projects.

The maximum grant available under the SGP is usually 50% of the total project costs. The Secretariat maintains and publishes a list of all projects approved by the Council still awaiting financial support.

Origins: The Parties at COP5 (Geneva, 1997) resolved to draw on funds from the UN Trust Fund reserve in order to finance projects considered worthy of support by the Scientific Council. US$142,000 was allocated for each year of the 1998-2000 triennium for conservation projects.

Key facts:
- The Programme started in 1997 at COP5.
- It has supported over 50 research and conservation projects.
- Over US$1.4 million has been distributed (typical awards range between US$20,000-US$30,000).
- Past recipients have included NGOs, foundations, research institutions and universities.

Appendix VI: Information Management and National Reporting

Effective and efficient conservation action requires information on which to base planning and decision-making. Implementation of the actions set out in Resolution 6.5, adopted in South Africa in 1999, has resulted in the development of the CMS Information Management System (CMS IMS). The CMS Information Management System provides access to the various components and services derived from the management and implementation of the CMS Information Management Plan.

This system was developed by The United Nations Environment Programme World Conservation Monitoring Centre (UNEP-WCMC) (Cambridge) on behalf of the CMS Secretariat. It brings together data from various expert organisations, knowledge generated within CMS and other biodiversity agreements as well as information provided by the Parties to CMS through their National Reports. This includes a synoptic overview of the information provided by the Parties in their reports, which is prepared for each COP in form of the document 'Analysis and Synthesis of National Reports'. The information, which has so far been integrated into the CMS Information Management System, is accessible through a search form by clicking on the IMS link under the section 'Species Activities' on the CMS website.

The CMS IMS offers the following categories of information:

a) Information about animals listed in the Appendices:
This option provides access to information available in Party Reports for Appendix I species regarding population size, trends, research and monitoring activities, etc. in conjunction with the information provided on-line by expert organisations, e.g. BirdLife, Fishbase, GROMS, IUCN Red listing, UNEP-WCMC.

b) Information about animal groups of special interest to the CMS:
This provides access to information available in Party Reports on animal groups of special interest to the Convention (e.g. Bats, Birds, Marine Turtles, Marine Mammals, and Terrestrial Mammals). It also covers issues such as implementing legislation, obstacles to migration, threats, impediments for action and assistance required.

c) Information about Parties to the CMS:
This option enables the user to obtain a country profile for each Party to the CMS, including: information on national agencies and non-governmental organisations relevant to the implementation of the convention; national implementation of resolutions and recommendations; mobilisation of financial and technical resources; use of technology (e.g. satellite telemetry); and species listed in the Appendices present in the country in question. This option also provides access to the CBD Clearinghouse Mechanism of each Party.

d) Information provided by Parties to the CMS on specific themes:
This option enables the user to attain a regional or world-wide overview across Parties on the current state of affairs on a number of themes relevant to the implementation of the Convention, such as implementation of resolutions and recommendations, mobilisation of financial and technical resources, and use of technology, e.g. satellite telemetry.

One of the main mandates of the Conference of the Parties (COP) is the review of the implementation of the Convention. Of particular importance are the review and assessment of the conservation status of migratory species and of progress made towards their conservation as well as the evaluation of progress made under specific Agreements. Based on the information received, reviewed and assessed, the COP is mandated to make Recommendations to the Parties for improving the conservation status of migratory species listed in CMS Appendices I and II.

To this end, the Secretariat is requested to obtain, from any appropriate source, reports and other information, which will further the objectives and reinforce the implementation of the Convention. It should then also arrange for the appropriate dissemination of this information.

As part of this information gathering function, under the Convention and Agreements, Parties are requested to inform the COP, through the Secretariat, at least six months prior to each ordinary meeting of the Conference, on measures they are taking to implement the provisions of the Convention. The National Reporting process is a key instrument to enable the COP to assess the overall status of the Convention, as it can assist with the consideration of lessons learned, formulate appropriate requests and guidance to subsidiary bodies and the Secretariat, and, most importantly, identify priorities for action.

The Convention on Migratory Species and its daughter Agreements, the Agreement on the Conservation of African-Eurasian Migratory Waterbirds (AEWA) and European Bats (Eurobats) propose to establish the first-ever global system of online national reporting (SONAR) for the family of Multilateral Environment Agreements (MEA). This project should be ready by 2008 and it will include all the CMS Memoranda of Understanding.

Online Reporting will allow:

- Improved information quality, consistency and transparency.
- Improved efficiency of information management and use for the Secretariats, Parties and the wider public.
- Reduced cost of information systems development for Secretariats (shared costs) and reporting costs for Parties.
- Improved linkages and sharing of information with international MEAs, environmental monitoring agencies, major data custodians, and regional treaties and the wider public.
- Reduced burden on national governments.

Appendix VII: The Global Register of Migratory Species (GROMS)

Background: Of the 1.5 million scientifically described animal species, 5,000 are estimated to be migratory. These are animals that cyclically migrate a minimum of 100 km and include 600 mammals, 2,000 birds, 10 reptiles, 1,000 fish and countless insects and other invertebrates. This number doubles to 10,000, if small-scale migrants which regularly cross national boundaries (the definition under CMS) are included. The number would further increase if geographically separate populations and subspecies exhibiting distinct migration patterns were individually registered.

The Global Register of Migratory Species (GROMS) supports the goals of the Convention by summarising our present knowledge of migratory species in a relational database connected to a geographical information system (GIS). The database provides fully referenced basic information on species, maps and range state lists and a bibliography in a flexible form which can be adapted to the user's needs. It contains scientific names with authority and synonyms; common names in English, French, German and Spanish; and species' status under CMS and its agreements, and their category according to IUCN International Red List and CITES Appendices.

The Zoological Museum in Bonn (the Museum Alexander Koenig) first initiated GROMS and attracted the support of the German Federal Environment Ministry (BMU) and the Federal Agency for Nature Conservation (BfN). The CMS Secretariat has assumed management responsibility.

GROMS has been uniquely tailored to store the diverse information on migratory animal species within one database, including maps. An electronic Geographic Information System (GIS) is connected to the database, allowing it to store the complexities of movements in space and time. This includes specific information about migration such as breeding, resting, hibernation areas and migration routes, which other databases do not have.

GROMS incorporates GIS maps for 700 species, including most on CMS Appendix I, and non-passerine bird maps adapted from the "Handbook of the Birds of the World" (del Hoyo et al., 1992-2000). Intersection with other GIS-layers is an efficient and professional way to analyse or predict the effects of habitat destruction, population density, land use or pollution on migratory species. GROMS is compatible with and provides links to other major species information systems, one such example being FishBase, a similar global database dedicated to fish species.

GBIF (the Global Biodiversity Information Fund) and CMS are working together under a Letter of Agreement to update and share the information stored in GROMS.

Key facts:

- GROMS, the geo-biological database for migratory species, contains information on: 4,344 vertebrate species, 5,600 references, 1,300 GIS-maps, 298 migratory mammals, 131 bats, 45 terrestrial mammals, 2,203 migratory birds, 10 reptiles (including 7 sea turtles), 1,926 migratory fish.

- GROMS can contribute to the implementation of the CMS Information Management Plan, provide information to Parties, scientific bodies and conservation organisations and maintain an updatable reference work.

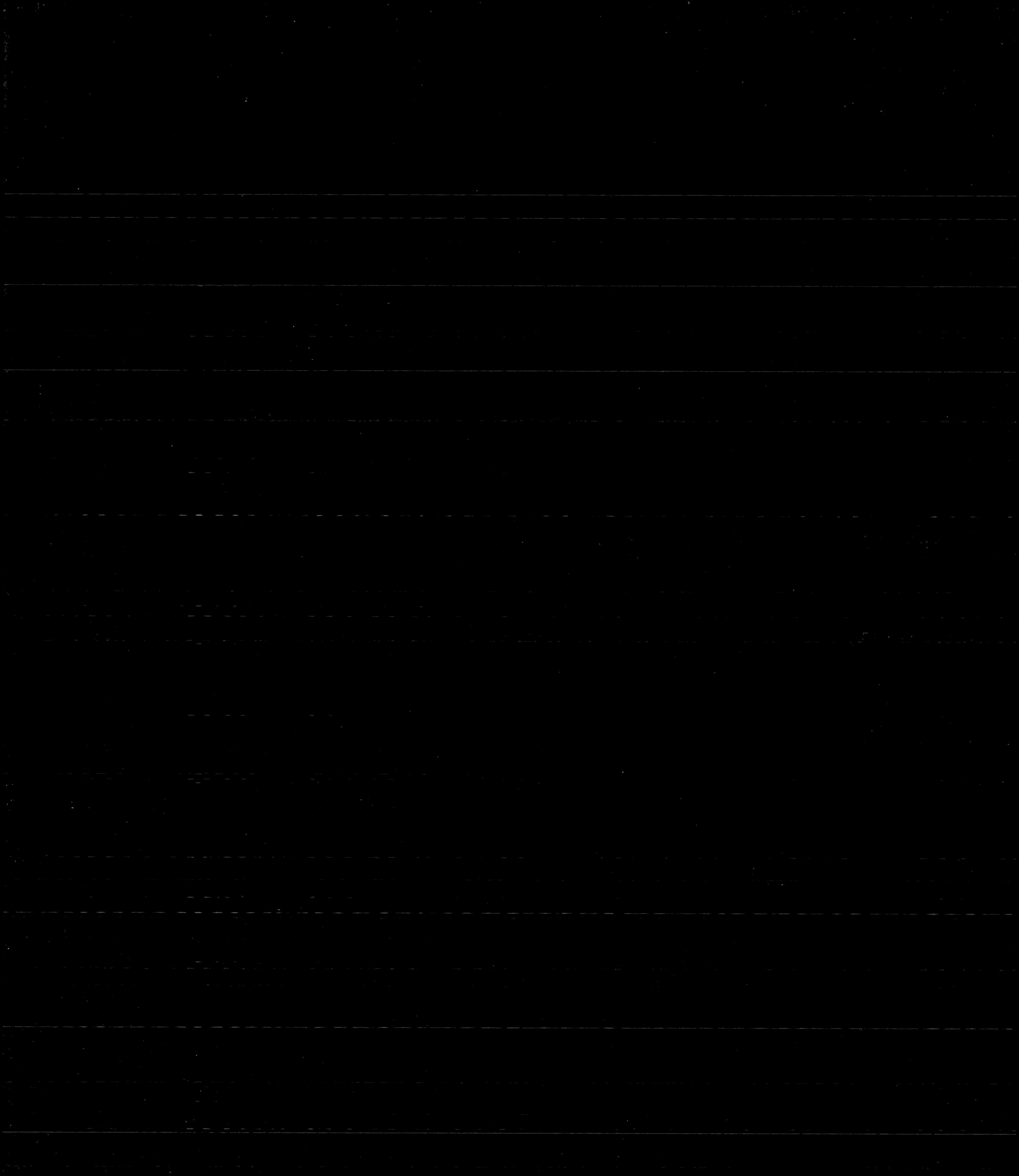

Photographic Credits

Prelims
1. Takako Uno/Splashdown/Still Pictures
2-3. Steven Kazlowski/Still Pictures
4-5. McPhotos/Still Pictures
6-7. NaturimBild/Foerster/Still Pictures
8-9. Wildlife /M.Hamblin/Still Pictures
10-11. Biosphoto/Crocetta Tony/Still Pictures
12-13. Biosphoto/Gunther Michel/Mau/Still Pictures
14. Biosphoto/Michel & Christine Denis-Huot/Still Pictures
16. Stanley Johnson
17. Biosphoto/Lorgnier Antoine/Still Pictures
18. Biosphoto/Michel & Christine Denis-Huot/Still Pictures
19. Biosphoto/Ruoso Cyril/Still Pictures
20. Biosphoto/Allofs Theo/Still Pictures
21. Biosphoto/Ruoso Cyril/Still Pictures
23. Tim Rock/WaterFrame/Still Pictures
24-25. Patrick Frischknecht/Still Pictures
26. H. Schulz/Still Pictures
28-29. Wildlife/S.Muller/Still Pictures

Chapter 1: Birds
30-31. Biosphoto/Mafart-Renodier Alain/Still Pictures
34-35. Biosphoto/Huguet Pierre/Still Pictures
36-37. Biosphoto/Stoelwinder/Still Pictures
38. Sergey Dereliev (UNEP/AEWA)
39. M. Woike/Still Pictures
40-41. Reinhard Dirscherl/WaterFrame/Still Pictures
42. Biosphoto/Chatagnon Jean-Paul/Still Pictures
43. McPhotos/Still Pictures
44. Wildlife/R.Usher/Still Pictures
45. M. Woike/Still Pictures
46. Wildlife/M.Lane/Still Pictures
47. A. Liedmann/Still Pictures
48-49. Jean-Leo Dugast/Still Pictures
50-51. Biosphoto/Watts Dave/Still Pictures
52. W. Wisniewski/Still Pictures
52-53. Biosphoto/Pons Alain/Still Pictures
54. Bill Coster/Still Pictures
55. Wildlife/J.M.Simon/Still Pictures
56-57. Biosphoto/Larrey Frédéric & Roger Thomas/Still Pictures
58-59. Wildlife/G.Delpho/Still Pictures

Chapter 2: Aquatic Species
60-61. Reinhard Dirscherl/WaterFrame/Still Pictures
63. Biosphoto/Vernay Pierre/Polar Lys/Still Pictures
64-65. Maren Reichelt/Sea Watch Foundation
66-67. Michael Nolan/Splashdown/Still Pictures
67. Danny Frank/Splashdown/Still Pictures
68-69. Biosphoto/Cole Brandon/Still Pictures
70-71. Jonathan Bird/Still Pictures
72-73. Kevin Schafer/Still Pictures
74-75. Andre Seale/Splashdown/Still Pictures
76. Wildlife/J.Freund/Still Pictures
76- 77. Biosphoto/Migeon Christophe/Still Pictures
78-79. A. Hartl/Still Pictures
80-81. Wildlife/J.Freund / Still Pictures
82-83. Doug Perrine/Still Pictures
84-85. Ingrid Visser/Splashdown/Still Pictures
86-87. Steven Kazlowski/Still Pictures

Chapter 3 : Terrestrial Species
88-89. McPhotos/Still Pictures
91. Biosphoto/Fouquet Franck/Still Pictures
92-93. Biosphoto/Gunther Michel/Still Pictures
94-95. Wildlife/I.Shpilenok/Still Pictures
95. Wildlife/I.Shpilenok/Still Pictures
96. Biosphoto/Courteau Christophe/Still Pictures
97. Martin Harvey/Still Pictures
98-99. Natalia Marmazinskaja/WWF Russia
100-101. McPhotos/Still Pictures
102-103. Biosphoto/Willocx Hugo/Wildlife Pictures/Still Pictures
104-105. Wildlife Pictures/Still Pictures
106-107. K. Wothe / Still Pictures
108-109. Steven Kazlowski/Still Pictures
110-111. McPhotos/Still Pictures
112-113. S. Meyers/Still Pictures
114-115. Biosphoto/Courteau Christophe/Still Pictures
116-117. Biosphoto/Michel Bureau/Still Pictures
118-119. Fischer/Alpaca/Andia.fr/Still Pictures
120-121. James L. Amos/Still Pictures

Chapter 4 : Threats & Challenges
122-123. Ricardo Funari/Still Pictures
125. C. Allan Morgan/Still Pictures
126-127. S. Zankl / Still Pictures
129. Andrew Stewart/Still Pictures
130-131. Biosphoto/Bily Guillaume/Still Pictures
132. Martin Harvey/Still Pictures
134-135. Ton Koene/Still Pictures
136-137. McPhotos/Still Pictures

Appendices
138-139. Doug Cheeseman/Still Pictures
158-159. Tim McCabe-UNEP/Still Pictures

33. map © The Stationary Office